A Guide to Systems Engineering and Management

A Guide to Systems Engineering and Management

Stanley M. Shinners
Sperry Division
Sperry Rand Corporation
and New York Institute of Technology

Lexington Books
D.C. Heath and Company
Lexington, Massachusetts
Toronto

Library of Congress Cataloging in Publication Data

Shinners, Stanley M.
  A guide to systems engineering and management.

  Includes index.
  1. Systems engineering. I. Title
TA168.S443   620'.7   76-7289
ISBN 0-669-00680-7

Copyright © 1976 by D.C. Heath and Company

All rights reserved. No part of this publication may be reproduced or transmitted in any form or by any means, electronic or mechanical, including photocopy, recording, or any information storage or retrieval system, without permission in writing from the publisher.

Published simultaneously in Canada

Printed in the United States of America

International Standard Book Number: 0-669-00680-7

Library of Congress Catalog Card Number: 76-7289

To my wife
Doris
and to my children
Sharon, Walter, and Daniel

Contents

	List of Figures	xi
	List of Tables	xvii
	Preface	xix
Chapter 1	**Scope of the Systems Engineering Problem**	1
	Introduction	1
	Overall Objectives of a System	3
	Feedback Characteristics of the System Problem	4
	Mathematical Representation of the System Problem	8
	Techniques for Systems Optimization	12
	Procedures for Engineering a System	14
Chapter 2	**Performance**	19
	Introduction	19
	Sources of System Error	20
	The Identification Problem	21
	Predictable and Unpredictable Errors	24
	Predictable Errors	24
	Characteristics of Unpredictable Errors	30
	Combining Errors	33
	The Error Budget	37
	An Example—A Tracking Radar System	37
Chapter 3	**System Reliability, Maintainability, and Availability**	45
	Introduction	45
	Definitions	46
	The Basic Foundations of a Reliability Program	48
	Use of Semiconductor Circuits for Improving Electronic System Reliability	51
	Redundancy	54
	Grouping and Failure Detection Techniques for Parallel Redundancy	55
	Probability Analysis of Group Redundancy	57
	Failure Detector Requirements	60

	Partial Redundancy	63
	Maintainability	66
	Reliability, Maintainability, and Availability (R/M/A)	67
Chapter 4	**Management Control of System Schedule and Cost**	69
	Introduction	69
	The Economic Flow Graph	70
	Program Management	72
	Time Models—PERT	74
	Mathematical Models of Engineering Organizations	86
	Program Performance—An Actual Case History	89
	Estimating Schedules and Costs	93
	Use of Computers as Aids for Controlling Schedule and Cost	95
Chapter 5	**Simulation**	99
	Introduction	99
	Analog Simulation	100
	Analog Computer Elements	101
	Analog Computer Simulation Techniques	106
	Digital Simulation	109
	Elements of Digital Computers	110
	Programming	112
	Hybrid Computers	118
	Monte Carlo Methods	120
	Random Signal Generators	125
Chapter 6	**Man-Machine Control Systems: The Human Model**	133
	Introduction	133
	Unique Characteristics of Human Controllers	134
	Linear, Continuous Models of the Human Controller	135
	Linear, Discrete Models of the Human Controller	139
	The Human Controller's Adaptive Capability	141
	Adaptive Representation of the Human Controller	146
	Techniques for Aiding the Human Controller	149
	Conclusions	152

Chapter 7	**Testing Techniques**		157
	Introduction		157
	The Basic Foundations of Systems Test		158
	Computer-Controlled Testing		159
	Computer Techniques for Analyzing Test Data		160
	Uncertainties Associated with System Test		163
	Example of Testing a Large Naval Shipboard System		165
Chapter 8	**Systems Management**		171
	Introduction		171
	Work Breakdown Structure		171
	Management Information Systems		172
	Systems Management Considerations		178
	Assumptions		178
	Alternatives		179
	Risk Assessment		180
	Communication Aspects of Systems Management		184
Chapter 9	**Problems**		189
	Index		215
	About the Author		223

List of Figures

1.1	Air Traffic Control System	2
1.2	Block Diagram Illustrating Procedure for Optimizing the System	5
1.3	Block Diagram Representation Illustrating Feedback Interrelationships of Overall System Performance, Reliability, Schedule, and Cost	6
1.4	Feedback Representation of the Systems Engineering Process	15
2.1	A Feedback Control System with an Undesired External Disturbance Input, $U(s)$	20
2.2	Identification Problem of an Open-Loop System	22
2.3	Identification Problem in a Closed-Loop System	22
2.4	An Adaptive System	23
2.5	Correcting the System Error Due to an External Disturbance in a Closed-Loop Manner	29
2.6	Correcting the System Output in an Open-Loop Manner	29
2.7	A Gaussian Probability Density Distribution	31
3.1	Typical Curve of Failure Rate vs. Time	47
3.2	A Two-Group Redundant System Having an Input to Both a Normal and a Redundant Channel with a Failure Detector across the Channel Outputs of Each Group	56
3.3	A Two-Group Redundant System Having an Input to Both a Normal and a Redundant Channel with a Failure Detector across Only the Normal Channel Output of Each Group	56
3.4	A Two-Group Redundant System with High-Frequency Test Signal Input to Both Normal and Redundant Channels	57
3.5	An N/n Group Redundant System Having an Input to Both a Normal and a Redundant Channel	58
3.6	A Simple System, Containing N Elements, Without the Use of Group Redundancy	59

3.7	Block Diagram of a Group Redundant System Illustrating the Use of Difference Amplifiers for Detecting Failures and Checking Which Channel has Failed	61
3.8	Comparator Failure Detector Using Pulse Modulation Techniques	62
3.9	Group Redundant Approach for Entire Feedback Control System	63
3.10	A Partially Redundant System	64
3.11	Availability as a Function of MTBF/MTTR	67
4.1	The Economic Flow Graph	71
4.2	Cost Feedback Path	73
4.3	Organization Chart of a Typical Systems Engineering Division of a Large Company	75
4.4	A General PERT Network	77
4.5	Probability Distribution Function: General Characteristics	78
4.6	Probability Distribution Function: Characteristic Variations	79
4.7	The Beta Probability Density Distribution	80
4.8	Comparison of Probability Distributions for Two Gaussian Time Functions Having Different Variances	82
4.9	PERT Diagram for Preparing Engineering Drawings to Manufacture a Special-Purpose Computer	84
4.10	Example for Determining "Slack" Time	85
4.11	Organizational Performance	87
4.12	Underdamped, Overdamped, and Critically Damped Organizational Responses	89
4.13	Mathematical Model Representation of an Engineering Department	90
4.14	Example of an Engineering Department's Response	91
4.15	Mathematical Models of Cost Curves	92
4.16	Block Diagram Representation of Program Illustrated in Figure 4.15	94
5.1	Basic Operational Amplifier Circuit	102
5.2	Simplified Representation of the Operational Amplifier	103
5.3	An Analog Adder	103

5.4	Relationship Between c, $B\dot{c}$, and $\ddot{c}$		107
5.5	Analog Computer Block Diagram Corresponding to Solution of the Differential Equation: $\ddot{c} + B\dot{c} + c = F$		107
5.6	Analog Computer Block Diagram for Solving the Simultaneous Equations: $\ddot{x} + \dot{x} + x + 2y = F$; $\ddot{y} + 4\dot{y} + 2y + 8x = 0$		108
5.7	On-line Computer Control		110
5.8	Off-line Computer Control		111
5.9	Block Diagram of a Typical Digital Computer		111
5.10	An Example of a Flow Diagram		114
5.11	Flow Chart for Computing $y = x^n/n!$		115
5.12	Detailed Flow Chart from Points a to a' of Figure 5.11		115
5.13	Detailed Flow Chart from Points a' to a'' of Figure 5.11		116
5.14	Analog Computer Simulation of the Split-Boundary Problem		119
5.15	Effect of Successive Trials in Achieving $c(2) = 1$		120
5.16	Automatic Hybrid Computer Determination of $\dot{c}(0)$		121
5.17	Block Diagram Representation of the Monte Carlo Method		122
5.18	Buffon's Needle Problem		123
5.19	A One-Dimensional Solution for Determining the Probability of Kill of a Missile System		124
5.20	A Random Signal Generator Utilizing a Radioactive Source		126
5.21	Output Waveform of a Radioactive Source		126
5.22	Statistical Characteristics of the Random Telegraph Signal		127
5.23	Conversion of the Random Telegraph Signal to a Low-Frequency Random Test Signal Having a Gaussian Distribution		128
5.24	A Feedback Shift Register		128
5.25	Autocorrelation Function for Shift Register Code		129
6.1	Pursuit Manual Tracking System		136
6.2	Compensatory Manual Tracking System		137
6.3	Ward's Discrete Model of the Human Controller as Redrawn by Bekey		140

6.4	Bekey's Discrete Model of the Human Controller	141
6.5	Various Types of Adaptive Characteristics Present in the Human Controller	143
6.6	History of a Pitch Damper Failure	145
6.7	Adaptive Model of Human Operator Proposed by McRuer and Krendel	147
6.8	"Model Reference" Adaptive Control System for the Human Controller	148
6.9	"Error Pattern Recognition Model" for the Human Controller	149
6.10	Aided Tracking Obtained by Adding the Human Controller's Output Directly to the Display	150
6.11	Aided Tracking Obtained by Adding the States of the Controlled System's Dynamics to the Display	151
6.12	A Rate-Aided Tracking System	152
6.13	A Possible Rate-Acceleration-Aided Tracking Configuration	153
7.1	Probability Relationships for Measuring an Unknown Value	164
7.2	Functional Flow Diagram of an Automatic Testing System	166
7.3	Major Features of Automatic Test Software	167
8.1	A Weapons System Work Breakdown Structure	173
8.2	The Development Process of an MIS	176
9.1	General Concept of a Closed-Loop Command and Control System	191
9.2	Block Diagram of a Linear Control System	192
9.3	Block Diagram for a Control System	192
9.4	Tokyo-to-Hakata Super Express Train	192
9.5	Block Diagram for Automatic Braking System of High-Speed Train	196
9.6	Submarine Depth Control Problem	197
9.7	Equivalent Block Diagram for Depth Control System of a Submarine	198
9.8	Three Computing Systems to Compare for Reliability	199

9.9	PERT Diagram for Preparing Engineering Drawings to Manufacture a Receiver	202
9.10	PERT Diagram for Preparing Engineering Drawings to Manufacture a Transmitter	203
9.11	Performance Graph for an Actual Program	204
9.12	Block Diagram Showing Relationship Between Wages, Prices and Cost of Living	204
9.13	Block Diagram of Process of Interest Accrual in a Savings Account	205
9.14	Block Diagram of National Economic Problems	206
9.15	Series Resonant Circuit	209

List of Tables

2.1	Summary of Steady-State Error Values for Various Types of Inputs	28
2.2	Relationship Between CPE and Orthogonal Normal Distributions	35
2.3	Error Budget for Elevation Axis of Tracking Radar	41
5.1	Summary of Basic Analog Computing Elements	105
5.2	Definition of Instruction Words	113
5.3	Program for Determining $y = x^n/n!$	117
8.1	Alternative Designs of a High-Performance Fighter Aircraft	180
8.2	Selection of Alternative High-Performance Fighter Aircraft Designs with the Dimension of Risk Added	180
8.3	A Task Risk Analysis Overview List: Weapons System Program, Tracking Radar Task	181
8.4	Guidelines for Assigning Risk Percentage	182
8.5	A Work Breakdown Risk Analysis Summary Sheet for Weapons System Program	183
8.6	Possible Checklist to be Used for Collecting Data for Determining Risk Assessment	185
9.1	Follow-On Contract Chart	214

Preface

The management of modern technologically complex programs requires individuals who are capable in many disciplines. These programs require a clear, unified understanding of various aspects of the problem in order to be solved in an effective manner. *A Guide to Systems Engineering and Management* provides a framework which can be applied to the challenging problems facing today's systems engineers and program managers in organizations of all sizes. In addition to assisting the practicing engineer, it should prove useful in systems engineering and management courses offered by industry and universities. An accompanying *Instructor's Guide and Solutions Manual* is available which provides a recommended course outline and detailed solutions to a set of practically oriented problems found at the end of the book.

The successful program manager must be a competent systems engineer, economist, and manager. As a systems engineer, he must understand analysis, simulation, modeling, reliability, and test techniques. In addition, he must be aware of state-of-the-art concepts and realize their limitations. As an economist and manager, the program manager must be able to recognize the risks of a program and know how to plan and control the fiscal and schedule aspects of the program. In addition, as a manager, he must be able to communicate with his superiors and subordinates in a manner which clearly and convincingly illustrates problems and their solutions.

A Guide to Systems Engineering and Management has been written to provide today's systems engineers and program managers with this capability. It should prove very useful to all practicing engineers desiring a unified and cohesive book which treats systems engineering and systems management from a practical and analytical viewpoint. In addition, it should enable students to relate theoretical concepts to practical problems. The book emphasizes the author's industrial experience as a program manager with Sperry Rand Corporation coupled with extensive teaching experience in industrial and university programs. The latest concepts, with practical examples, are included in this unified treatment.

The systems engineer must have an adaptive capability which distinguishes him from the engineering specialist who is concerned with only one aspect of a well-defined engineering discipline. The systems engineer is expected to adapt to the requirements of almost every type of system such as transportation, energy, control, and management. Although particular systems differ in their details, they usually have common structural relations which can be recognized and utilized by the systems engineer. The objectives of a system are considered from the viewpoints of performance, reliability, schedule, cost, maintainability, power consumption, weight, and life expectancy.

In addition, *A Guide to Systems Engineering and Management* illustrates how the interelationships among these various systems parameters can be viewed. After presenting basic systems concepts, each chapter illustrates the application of these techniques to practical systems engineering problems. These include systems such as air traffic, data processing, and automatic control systems. A presentation of the latest techniques for systems management is also included. The final chapter contains practical engineering problems to supplement each of the preceding chapters.

Chapter 1 introduces the nature of the problem and basic systems concepts; emphasizes the feedback characteristics of the systems engineering problem; and outlines the general procedure to be followed for designing a large complex system. Systems performance concepts are presented in Chapter 2 where predictable and unpredictable errors are considered and methods for analyzing them are presented. Chapter 3 discusses methods for predicting system reliability and techniques for improving maintainability and availability. Chapter 4 is concerned with management control of system schedule and cost by the utilization of such time models as PERT, mathematical models of an engineering organization's performance with actual examples, and computers. Methods for simulating systems are discussed in Chapter 5 from the viewpoints of analog, digital, and hybrid computers. Chapter 6 emphasizes the man-machine relationship in modern complex systems and presents several models, pertaining to human beings, which are useful to the systems engineer. Testing techniques are discussed in Chapter 7 where computer methods for performing a system test and analyzing test data are emphasized. These concepts are then applied to the testing of a large system. Chapter 8 presents several modern methods used in the management of large, complex systems. These include such techniques as work breakdown structures, management information systems, and risk assessments.

Examples of practical problems, chosen specifically to enhance the presentation of each chapter and to emphasize salient points, conclude the text in Chapter 9. This is a helpful feature of the book for the practicing engineer who desires to use the book for self-study, and for the student. In addition, detailed solutions to the problems are contained in an accompanying *Instructor's Guide and Solutions Manual*.

I express my sincere appreciation to Dr. Thomas J. Higgins, Professor of Electrical and Computer Engineering at the University of Wisconsin, Madison, for his very useful comments and suggestions. I thank *Machine Design* for granting the permission to use portions of my paper "Which

Computer . . . Analog, Digital, or Hybrid?'' (January 21, 1971) in Chapter 5, "Simulation." I also thank *Computers & Electrical Engineering* for permission to use portions of my paper, "Considerations for Automatic Testing of Complex Shipboard Systems" (June 1973, Vol. 1, No. 1) in Chapter 7, "Testing Techniques".

I am most grateful to my wife Doris and daughter Sharon Rose for their encouragement, understanding, patience, and typing assistance throughout. In addition, I express thanks and appreciation to my parents for their efforts, encouragement, and inspiration.

Stanley M. Shinners

Jericho, New York
February 1976

A Guide to Systems Engineering and Management

1 Scope of the Systems Engineering Problem

Introduction

Modern systems are becoming increasingly larger and more complex. The existence of these large systems and their continuing growth is due to the increasing size and complexity of modern industrial, military, aerospace, and business requirements. When large numbers of control, computer, and communications subsystems are interconnected to form a complex system, special attention must be given to the interaction of these subsystems in addition to their individual characteristics.

A large complex system consisting of control, computer, and communications systems, functioning in a highly integrated and interdependent manner to achieve overall sound performance, reliability, schedule, cost, maintainability, power consumption, weight, and life expentancy, is referred to as the *systems engineering problem*.[1,2] Although systems engineering is not an entirely new subject, its scope and the engineering techniques utilized have undergone rapid and extensive developments recently. This is especially true in such diverse fields as power generation, police and fire command and control systems, space vehicle control, air traffic control, ground traffic control, high-speed rail transportation systems, and military weapons systems.

Large modern complex systems are characterized by control systems, computers, and communications networks functioning in a highly integrated and interdependent manner. For example, consider the modern air traffic control system for a large metropolitan airport (Figure 1.1). This system is a closed-loop system which has the ability to control the air traffic environment based on information presented to the air traffic controller. These decisions are based on information obtained by radar, communicated by a data link to a computer for data processing, and appropriately displayed to the air traffic controller who then closes the loop by communicating commands to the pilots of the aircraft to control the overall air traffic environment. The feedback loop continues to operate until the air traffic environment is exactly what the air traffic controller desires it to be.

Observe from this example that the human element is an important part of this system in addition to the control, computer, and communication elements. Due to the ever-increasing volume and speed of modern aircraft, the human decision element cannot integrate all of the changing dynamic

1

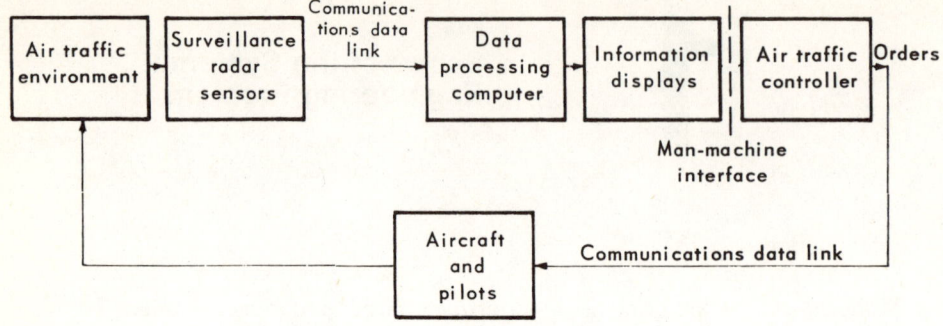

Figure 1.1. Air Traffic Control System

data to formulate decisions which will result in the smoothest, safest use of air traffic space. Modern air traffic control systems must rely greatly on automatic predictions from computers concerning various alternatives for contingency conditions. To accomplish this, a large volume of data must be collected, collated, processed, and organized quickly, accurately, and reliably.

The air traffic control problem demonstrates that modern, large, complex systems are in reality man-machine control systems. Although great emphasis is placed on the design of elaborate equipment, the final performance is dependent on the interaction of man and machines operating in a highly integrated, unified manner. A very practical systems engineering consideration is to determine the boundaries for separating the machine from the human-decision element. Great advances have been made in self-organizing machines and adaptive learning machines.[3,4] These developments, which have resulted in a better understanding of the integrated behavior of man and machine, are a very important part of systems engineering considerations.

Since most modern, complex systems are dependent on feedback, this book emphasizes and focuses special attention on systems engineering aspects in light of feedback system concepts. Control theory alone does not solve the entire engineering problem. However, it is a very important systems engineering tool. The control engineer is in a unique position to become a systems engineer because of his basic knowledge of block diagrams, feedback control theory, computers, simulation, and modeling. In addition, systems engineers must be knowledgeable in the area of overall systems performance, reliability, economics, optimization, testing, and how to treat a human being in the loop to synthesize and evaluate an entire complex system.

The aim of this book is to discuss the overall aspects of composite systems and serve as a catalyst to the reader, who has some background in basic feedback system concepts, to enable him to engineer large, complex, modern systems. Basic concepts regarding systems performance are discussed in Chapter 2; reliability and maintainability in Chapter 3; schedule and cost in Chapter 4; systems simulation is presented in Chapter 5; the characterizations of the human element are developed in Chapter 6; and system testing techniques are presented in Chapter 7. Chapter 8 illustrates the techniques for managing a system. These are the ingredients of the systems engineering process. The systems engineer and program manager must consider all these factors in the evolving of the system.

The aim of this chapter is to examine the overall systems engineering problem; the following chapters examine the various elements involved in greater detail.

Overall Objectives of a System

The overall objectives of a particular system are very complex which vary widely and depend on specific requirements. Simple one-word answers such as performance, reliability, or cost alone are not sufficient. Actually, all these factors must be considered in the design of any system. Other factors usually considered are schedule, maintainability, power consumption, weight, and life expectancy.

Depending upon the application and the customer's goals, some systems may stress performance, others reliability, others schedule, and others cost. For example, the design of one high-performance positioning system for a land-based instrumentation tracking radar used for communicating with a space satellite, would stress performance and reliability, and place less emphasis on the importance of cost and weight. However, the design of a commercial hydraulic positioning system used in road construction, which is to be manufactured in very large quantities, would stress cost and relegate the importance of reliability and performance to lesser roles. Because of these basically differing requirements, it is necessary to consider each system separately and judge the relative importance of the various objectives in their proper perspective.

A modern systems engineer may view optimization of an overall unified system via the general block diagram illustrated in Figure 1.2. The inputs to the system are the system requirements, resources in the form of manpower, equipment and materials, and the environment which includes the available power sources, operating temperature, humidity and external disturbances. Utilizing an appropriate weighting function for each desired system characteristic, a set of weighted system functions are obtained which can be used to optimize the system functions. The appropriate weighting functions for performance, reliability, schedule, cost, power consumption, weight, maintainability, and life expectancy, are denoted by W_P, W_R, W_S, W_C, W_{PC}, W_W, W_M and W_L, respectively. The system specification can be formulated from the output of the process illustrated in Figure 1.2, and denoted as the "overall optimum system function," W_T.

The values of the weighting functions depend on the particular application of each system. Their relative values determine the resulting overall system to a major degree. The weighting functions can be made to have the adaptive characteristic of time-variableness. For example, after the desired system performance has been achieved, W_P may be decreased and more importance can be placed on the cost by increasing W_C. In another realistic situation, consider a system where schedule has not been maintained. For this case, W_S would have to be increased and its overall effect distributed in some linear or nonlinear manner to the other weighting functions.

Feedback Characteristics of the System Problem

System performance, reliability, schedule, cost, maintainability, power consumption, weight, and life expectancy are interrelated system functions. For example, improved system performance can usually be obtained if the schedule and cost factors are increased. On the other hand, the cost of the system can be decreased at the expense of poorer performance and reliability. This interrelationship implies that the overall system can be considered from the viewpoint of multiple feedback paths, where several system characteristics are common to more than one feedback path. Decisions made to achieve one system objective usually also affect others, and the systems engineer must properly view and understand this feedback relationship.

The block diagram of Figure 1.3 is an interesting way to view this interrelationship.[5] In the *performance feedback path*, the desired overall system performance is compared to the anticipated overall system performance. The main elements of this closed-loop path are the specified overall system requirements, specified characteristics of the subsystems, determination of the overall system performance equations, and the calculation of the resulting overall systems' performance. The elements denoting specified overall system requirements and specified characteristics of the subsystems are also common to the *cost and schedule feedback paths*. In these closed loops, desired cost is compared to anticipated cost, and desired schedule is compared to anticipated schedule. Logic elements are utilized in these closed-loop systems for varying the specified overall system requirements and subsystem characteristics, and determining their

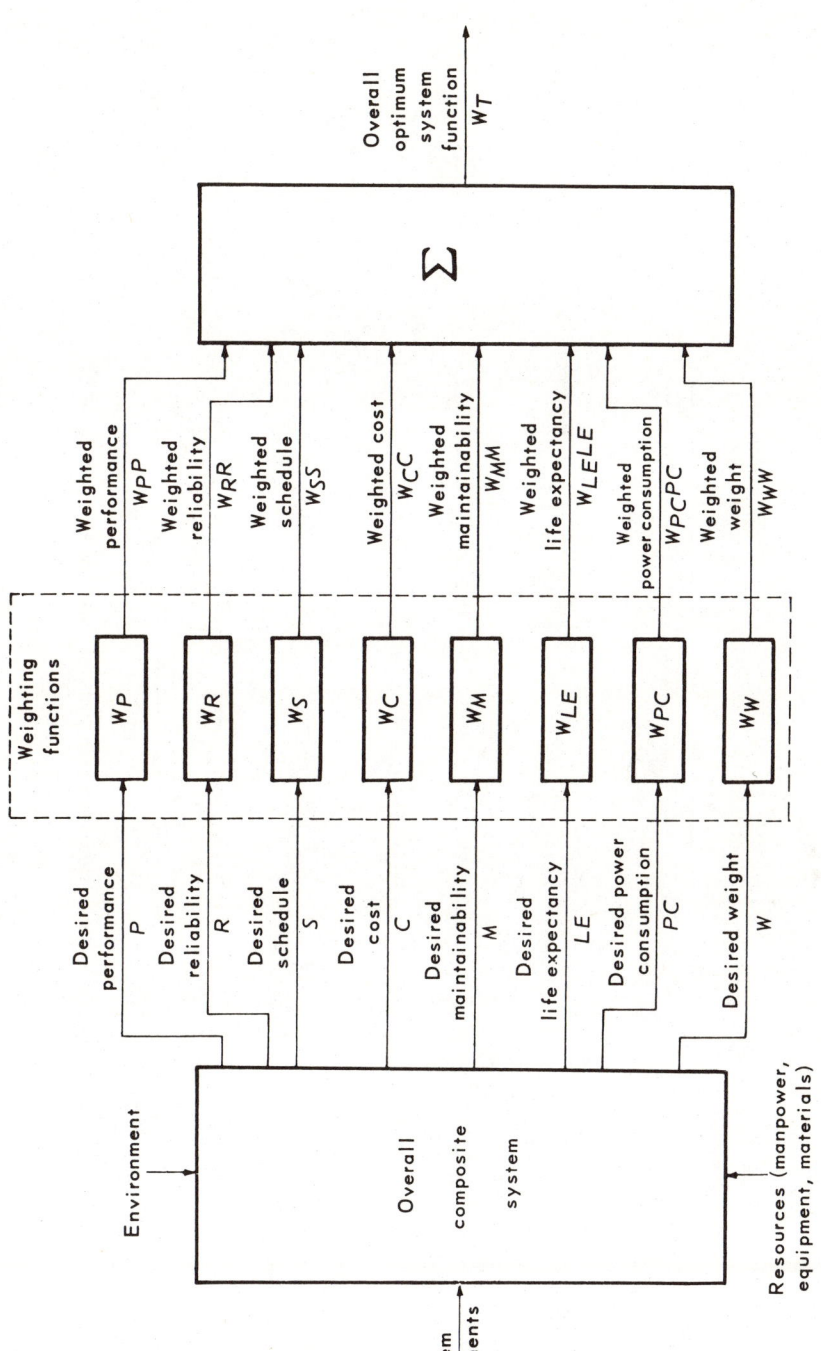

Figure 1.2. Block Diagram Illustrating Procedure for Optimizing the System

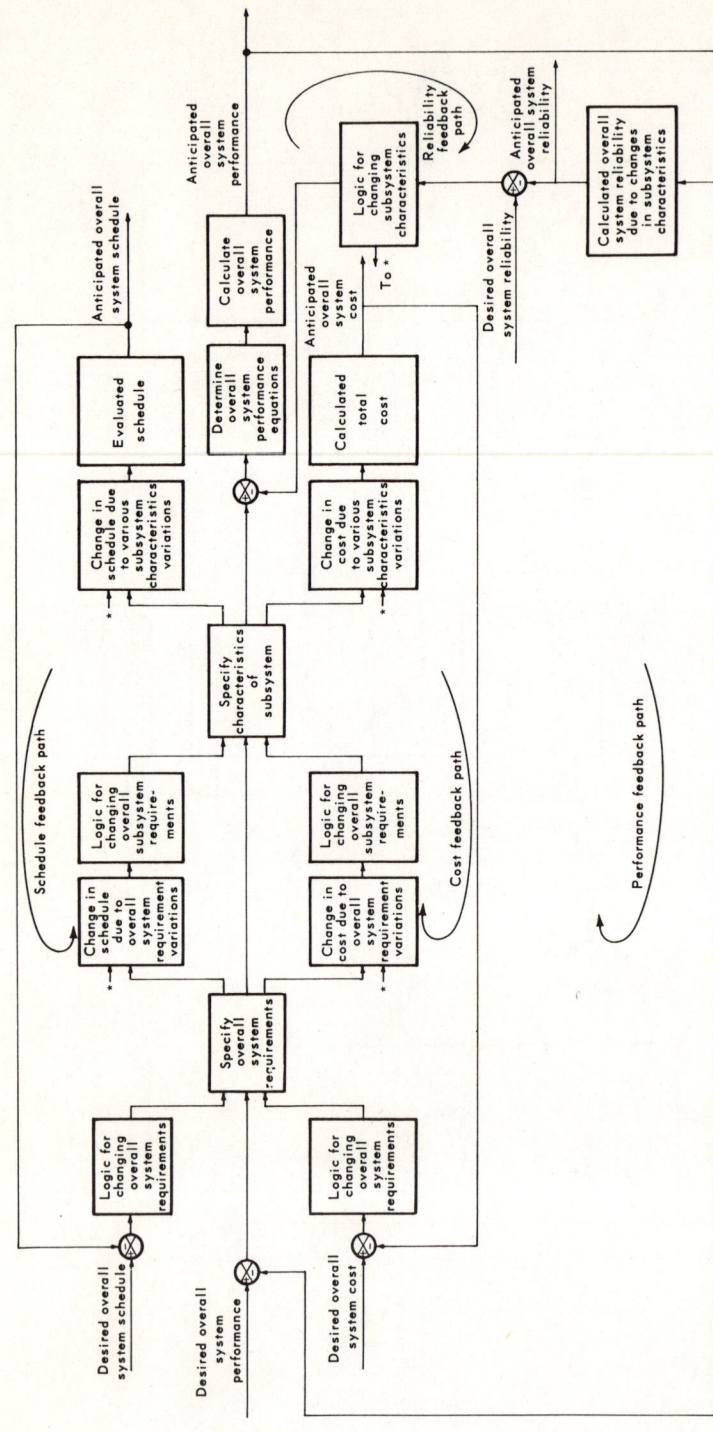

Figure 1.3. Block Diagram Representation Illustrating Feedback Interrelationships of Overall System Performance, Reliability, Schedule, and Cost

overall effect on cost and schedule. The elements of the performance feedback path which are concerned with determining the overall system performance equations and the calculated overall system performance, are also common to the *reliability feedback path* where desired overall system reliability is compared to the anticipated control system reliability. The element concerned with the calculation of the overall system relationship due to changes in the subsystem's characteristics takes into account the anticipated environment, materials, and the probability of changes occurring.

Study of this simplified block diagram demonstrates that several elements are common to more than one loop. Therefore, changes in performance, cost, schedule, and reliability are interrelated. This block diagram feedback concept can be extended to include maintainability, power consumption, weight, and life expectancy.

The elements indicated in the feedback loops of Figure 1.3 do not necessarily imply linear functions. In general, they may be time variable, nonlinear, and discrete components. In addition, some may behave in a statistical manner. If the element is nonlinear, for example, superposition is not valid and the effect of this element on all the feedback paths may not be the same.

The systems engineering problem is inherently feedback in nature. An engineered system is the result of compromise among such factors as performance, reliability, cost, schedule, maintainability, power consumption, weight, and life expectancy. Approximations and estimations play a very important role in the overall evolution of a unified system. Therefore, it is most important that the systems engineer responsible for the program be knowledgeable, experienced, and objective. In addition, due to the complexity of system problems, a digital computer can be a great asset in evaluating the overall problem.

As the system evolves, additional and better data are made available which will undoubtedly change the initial program plan. Therefore, it is wise to program the overall interrelationships, illustrated in Figure 1.3, on a computer and update the model with additional and new data. This procedure will save time and money in the long run, rather than doing the calculations over each time by hand. It is obvious that the initially conceived overall system is the best system based upon the data available at that time. An astute systems engineer must realize that his original model must be continually updated with additional and better data in the future. The result is that certain system parameters will change. Hopefully, this will reduce cost and schedule while improving performance and reliability. However, if the original estimates and approximations were made on a poor foundation, the opposite may occur.

Mathematical Representation of the System Problem

To illustrate the quantitative characteristics of the systems problem, let us consider the air traffic control system discussed previously and was illustrated in Figure 1.1. Based on specific systems requirements regarding performance, reliability, schedule, cost, maintainability, power consumption, weight, and life expectancy, it is desirable to formulate the problem mathematically to determine the optimum system solution.

The system should be engineered with respect to specific *performance* requirements. For example, the air traffic control system must be able to control a specific number of targets flying at a specified maximum speed and performing a specified maximum acceleration. Their position and velocity vectors must be determined with a certain specified accuracy. Furthermore, the system must be able to sense aircraft in the overall air traffic environment with a specified probability of detection. The significant performance parameters are designated:

P_1 = Target capacity

P_2 = Maximum target speed

P_3 = Maximum target acceleration

P_4 = Accuracy of position vector

P_5 = Accuracy of velocity vector

P_6 = Probability of detection. (1.1)

The *reliability* of the air traffic control system is extremely important. The overall system must be designed to result in the safe control of each aircraft from origin to destination. The overall system reliability is specified to be R, where

$$R = \text{Hours of mean-time-between-failures.} \quad (1.2)$$

The *schedule* for designing, fabricating, testing, and installing the entire system is a very important system factor. Adherence to these goals ultimately will determine the success of the program. The schedule objectives are:

S_1 = Design schedule

S_2 = Fabrication schedule

S_3 = Testing schedule

S_4 = Installation schedule. (1.3)

System *cost* objectives can be viewed in terms of the initial cost of the equipment, the maintenance cost, and the operational cost. The cost of maintenance and operation of the system over a specified number of years is an extremely important consideration. The cost objectives are specified:

C_1 = Initial cost of system

C_2 = Cost of maintenance over a ten-year period (1.4)

C_3 = Operational cost over a ten-year period.

The *maintainability* of the air traffic control system is a highly significant factor since it specifies the mean downtime after system failure. It is a very small and critical number since we are concerned with the large commercial airport which operates on a continuous basis. System maintainability is specified to be

$$M = \text{Minutes of downtime after failure.} \quad (1.5)$$

System *power consumption, weight*, and *life expectancy* are less critical numbers. However, the system must also be designed to satisfy specifications on these factors. System requirements concerning power consumption (*PC*), weight (*W*), and life expectancy (*LE*) are specified:

PC = X kilowatts

W = Y pounds (1.6)

LE = Z years.

The dependent system performance parameters and other factors represented by Equations (1.1) through (1.6) are functions of the aircraft surveillance radar sensors, data processing computer, information displays, the air traffic controller, the pilots, and the communications data links. Uncertainties due to weather, ground conditions, ceiling, and the motivation of the human elements in the system also affect these functions. Let us represent these independent factors as follows:

A = Aircraft

B = Surveillance radar sensors

D = Data processing computer

E = Information displays

F = Air traffic controller (1.7)

G = Pilots

L = Communication Data Links

U = Uncertainty factors.

These factors may be characterized by linear, nonlinear, time invariant, time variable, discrete, and/or statistical functions.

Equations (1.1) through (1.6) can be represented as functions of the various factors listed in Equation (1.7). For example, performance measured in terms of target capacity is a function of all the factors enumerated in Equation (1.7). Therefore, we can say that

$$P_1 = f_{P_1}(A, B, D, E, F, G, L, U). \tag{1.8}$$

It is important to emphasize that this functional relationship can be linear, nonlinear, time invariant, time variable, discrete and/or statistical. In a similar manner, the following expressions are obtained:

$$P_2 = f_{P_2}(A, B, D, E, F, G, L, U) \tag{1.9}$$

$$P_3 = f_{P_3}(A, B, D, E, F, G, L, U) \tag{1.10}$$

$$P_4 = f_{P_4}(B, D, E, F, L) \tag{1.11}$$

$$P_5 = f_{P_5}(B, D, E, F, L) \tag{1.12}$$

$$P_6 = f_{P_6}(B, D, E, F, L, U) \tag{1.13}$$

$$R = f_R(B, D, E, F, L, U) \tag{1.14}$$

$$S_1 = f_{S_1}(B, D, E, L) \tag{1.15}$$

$$S_2 = f_{S_2}(B, D, E, L) \tag{1.16}$$

$$S_3 = f_{S_3}(B, D, E, L) \tag{1.17}$$

$$S_4 = f_{S_4}(B, D, E, L) \tag{1.18}$$

$$C_1 = f_{C_1}(B, D, E, L) \tag{1.19}$$

$$C_2 = f_{C_2}(B, D, E, L, U) \tag{1.20}$$

$$C_3 = f_{C_3}(B, D, E, F, L, U) \tag{1.21}$$

$$M = f_M(B, D, E, F, L, U) \tag{1.22}$$

$$PC = f_{PC}(B, D, E, L, U) \tag{1.23}$$

$$W = f_W(B, D, E, L) \tag{1.24}$$

$$LE = f_{LE}(B, D, E, L, U). \tag{1.25}$$

Equations (1.8) through (1.25) can be rewritten in terms of functional coefficients. For example, P_1 given by Equation (1.8) may be a linear function of its factors as follows:

$$P_1 = N_{P_1A}A + N_{P_1B}B + N_{P_1D}D + N_{P_1E}E + N_{P_1F}F + N_{P_1G}G$$
$$+ N_{P_1L}L + N_{P_1U}U \tag{1.26}$$

The functional coefficients, represented by N_{xy} can be time-variable, non-linear, discrete, and/or statistical. In a similar manner, Equations (1.9) through (1.25) can be represented as follows:

$$P_2 = N_{P_2A}A + N_{P_2B}B + N_{P_2D}D + N_{P_2E}E + N_{P_2F}F + N_{P_2G}G + N_{P_2L}L + N_{P_2U}U \tag{1.27}$$

$$P_3 = N_{P_3A}A + N_{P_3B}B + N_{P_3D}D + N_{P_3E}E + N_{P_3F}F + N_{P_3G}G + N_{P_3L}L + N_{P_3U}U \tag{1.28}$$

$$P_4 = N_{P_4B}B + N_{P_4D}D + N_{P_4E}E + N_{P_4F}F + N_{P_4L}L \tag{1.29}$$

$$P_5 = N_{P_5B}B + N_{P_5D}D + N_{P_5E}E + N_{P_5F}F + N_{P_5L}L \tag{1.30}$$

$$P_6 = N_{P_6B}B + N_{P_6D}D + N_{P_6E}E + N_{P_6F}F + N_{P_6L}L + N_{P_6U}U \tag{1.31}$$

$$R = N_{RB}B + N_{RD}D + N_{RE}E + N_{RF}F + N_{RL}L + N_{RU}U \tag{1.32}$$

$$S_1 = N_{S_1B}B + N_{S_1D}D + N_{S_1E}E + N_{S_1L}L \tag{1.33}$$

$$S_2 = N_{S_2B}B + N_{S_2D}D + N_{S_2E}E + N_{S_2L}L \tag{1.34}$$

$$S_3 = N_{S_3B}B + N_{S_3D}D + N_{S_3E}E + N_{S_3L}L \tag{1.35}$$

$$S_4 = N_{S_4B}B + N_{S_4D}D + N_{S_4E}E + N_{S_4L}L \tag{1.36}$$

$$C_1 = N_{C_1B}B + N_{C_1D}D + N_{C_1E}E + N_{C_1L}L \tag{1.37}$$

$$C_2 = N_{C_2B}B + N_{C_2D}D + N_{C_2E}E + N_{C_2L}L + N_{C_2U}U \tag{1.38}$$

$$C_3 = N_{C_3B}B + N_{C_3D}D + N_{C_3E}E + N_{C_3F}F + N_{C_3L}L + N_{C_3U}U \tag{1.39}$$

$$M = N_{MB}B + N_{MD}D + N_{ME}E + N_{MF}F + N_{ML}L + N_{MU}U \tag{1.40}$$

$$PC = N_{(PC)B}B + N_{(PC)D}D + N_{(PC)E}E + N_{(PC)L}L + N_{(PC)U}U \tag{1.41}$$

$$W = N_{WB}B + N_{WD}D + N_{WE}E + N_{WL}L \tag{1.42}$$

$$LE = N_{(LE)B}B + N_{(LE)D}D + N_{(LE)E}E + N_{(LE)L}L + N_{(LE)U}U. \tag{1.43}$$

The resulting Equations (1.26) through (1.43) represent the system performance parameters expressed as a linear combination of factors in terms of their function coefficients.

The overall optimum system function, W_T, is the sum of the weighted functions of all the system objectives. From Figure 1.2, we obtain the following relationship:

$$W_T = W_P P + W_R R + W_S S + W_C C + W_M M + W_{(PC)}(PC) + W_W W + W_{(LE)}(LE). \tag{1.44}$$

Separating performance, schedule, and cost into the various components specified for this problem, this expression can be expanded:

$$W_T = W_{P_1}P_1 + W_{P_2}P_2 + W_{P_3}P_3 + W_{P_4}P_4 + W_{P_5}P_5$$
$$+ W_{P_6}P_6 + W_R R + W_{S_1}S_1 + W_{S_2}S_2 + W_{S_3}S_3$$
$$+ W_{S_4}S_4 + W_{C_1}C_1 + W_{C_2}C_2 + W_{C_3}C_3 + W_M M$$
$$+ W_{(PC)}(PC) + W_W W + W_{(LE)}(LE). \qquad (1.45)$$

The overall optimum system function, W_T, can be expressed in terms of the individual functional coefficients N_{xy} by substituting from Equations (1.26) through (1.43) into Equation (1.45). The result is:

$$W_T = [N_{P_1A}W_{P_1} + N_{P_2A}W_{P_2} + N_{P_3A}W_{P_3}]A$$
$$+ \left\{ \sum_{X=B,D,E,L} [N_{P_1X}W_{P_1} + N_{P_2X}W_{P_2} + N_{P_3X}W_{P_3} \right.$$
$$+ N_{P_4X}W_{P_4} + N_{P_5X}W_{P_5} + N_{P_6X}W_{P_6}$$
$$+ N_{RX}W_R + N_{S_1X}W_{S_1} + N_{S_2X}W_{S_2}$$
$$+ N_{S_3X}W_{S_3} + N_{S_4X}W_{S_4} + N_{C_1X}W_{C_1}$$
$$+ N_{C_2X}W_{C_2} + N_{C_3X}W_{C_3} + N_{MX}W_M$$
$$\left. + N_{(PC)X}W_{PC} + N_{WX}W_W + N_{(LE)X}W_{LE}]X \right\}$$
$$+ [N_{P_1F}W_{P_1} + N_{P_2F}W_{P_2} + N_{P_3F}W_{P_3} + N_{P_4F}W_{P_4} + N_{P_5F}W_{P_5}$$
$$+ N_{P_6F}W_{P_6} + N_{RF}W_R + N_{C_3F}W_{C_3} + N_{MF}W_M]F$$
$$+ [N_{P_1G}W_{P_1} + N_{P_2G}W_{P_2} + N_{P_3G}W_{P_3}]G$$
$$+ [N_{P_1U}W_{P_1} + N_{P_2U}W_{P_2} + N_{P_3U}W_{P_3} + N_{P_6U}W_{P_6} + N_{RU}W_R$$
$$+ N_{C_2U}W_{C_2} + N_{C_3U}W_{C_3} + N_{MU}W_M + N_{(PC)U}W_{PC} + N_{(LE)U}W_{LE}]U.$$

A meaningful analysis of Equation (1.46) could be accomplished only by utilizing a computer.

This mathematical representation of the system engineering problem is useful for identifying the complexity of the relationships between system objectives and system parameters. In addition, it yields guidelines for developing a logic which can result in a system design that is self-optimizing in order to obtain the best system design.

Techniques for System Optimization

The system problem formulated mathematically in the preceding section indicates that system optimization is a highly complex problem requiring a computer for solution. In addition, several simultaneous equations indicate that matrix techniques are quite useful. Developing a model of the system parameters and analyzing it on a digital computer is a very powerful tool for the systems engineer. However, it is still very important to formulate the problem correctly before utilizing this very powerful approach.

The task of assigning proper values to the system weighting functions is a very important and complex process. In some respects, performance, reliability, schedule, cost, maintainability, power consumption, weight, and life expectancy are all factors which must be considered in any system. Assigning relative weighting functions to these system objectives involves considerable experience. Trade-offs at the preliminary engineering level are extremely important and useful.

Trade-offs involve the comparison of reliability, schedule, and cost for different values of system performance. This process involves human judgment and depends greatly on the specific systems engineer's experience, knowledge, likes, and dislikes. It is subject to the human frailties of lack of good judgment and prejudices. It is not very unusual for two experienced systems engineers, with similar backgrounds, to disagree on what course of action should be followed, given the same set of facts and supporting data.

Computer modeling techniques, applied to the problem of determining relative system parameters, yield very useful results. The systems engineer can formulate the problem in which a system having multiple interacting inputs and outputs is best designed with respect to a certain performance criterion and known system limitations.

Comparing the set of solutions available to the systems engineer, the techniques of modeling provide answers to the problem which are less controversial than those obtained from trade-offs. However, these techniques usually require a digital computer for solution, and that is costly.

Trade-offs depend primarily on human judgment and can be performed relatively inexpensively given the facts and supporting data. However, the human frailties of poor judgment and prejudices cloud the conclusions reached from trade-offs while there can be very little argument about the answer to an analytical problem. Thus, the modern systems engineer has a decision to make on this point: Should he use trade-off techniques which are relatively inexpensive to perform but are subject to different interpretations, or should he use analytic solutions which are relatively costly to perform but which result in very little controversy? The problem should be viewed as a two-stage process. First, the method of trade-offs should be applied in order to get a coarse "ballpark" indication of the answer. This step may also point out certain problem areas which may have not been evident before, and may result in the systems engineer requesting more and/or better data. After this coarse solution has been obtained using trade-offs, the second step involves setting the model on a computer to obtain a fine, analytic solution to the problem.

This two-stage process has much merit. The engineer always should try to get a "ballpark" answer to a problem before he puts it on a computer. This initial step serves the following purposes:

1. To survey the overall problem and determine any problem areas which may have not been evident before. This may result in the engineer's requesting more and/or better data.

2. To serve as a check of the computer solution.

Following this procedure, the systems engineer will determine the system parameters in a very straightforward, logical manner. The objections to human frailties will be eliminated, and the advantages of human judgment and computer solution feedback correlation will be accomplished.

Procedures for Engineering a System[6,7]

The process involved in engineering a large complex system is a logical sequence of activities and decisions which lead to the definition, development, construction, and operation of the configuration. This process is conducted from the initial concept concerning the system objectives and is continued throughout the process for updating and reassuring necessary requirement changes. Systems engineering, therefore, is a closed-loop, iterative process as illustrated in Figure 1.4. The closed loops must feed the design solutions back to the system to determine whether the original set of requirements and theoretical predictions have been met. As illustrated in Figure 1.4, additional feedback paths result from a comparison of test data with the system objectives and requirements.

The initial stage of the systems engineering process involves a *conceptual analysis* of the system objectives. The systems engineer must fully determine the objectives, and at the same time, fully understand the real-world constraints imposed. Simulations and feasibility studies are performed at this stage.

The second step, *mission and requirements analyses*, also involves simulation. This step takes into account such factors as performance, reliability, maintainability, schedule, cost, power consumption, weight and life expectancy. After considering the factors in the mission simulation, and examining the operating environment of the equipment, the system's mission and requirements can be defined.

Functional analyses are performed in the third step. System and subsystem functions are analyzed in this step on the basis of the established set of system requirements and the time frame in which they must be accomplished. For example, in the design of a high-performance aircraft, some of the system functions to be considered would be weapons delivery and

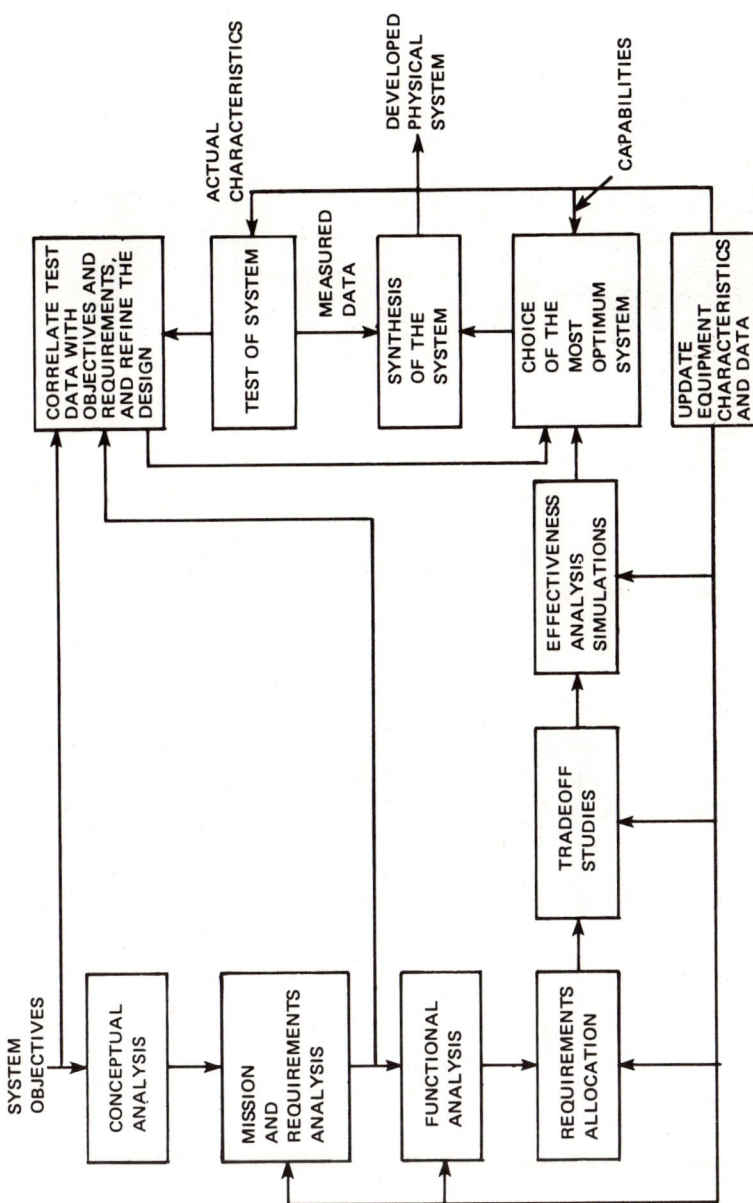

Figure 1.4. Feedback Representation of the Systems Engineering Process

air-to-air combat; some of the subsystem functions to be considered would be radar, navigation, weapons, display/control and communications subsystems. Each function of the system and subsystem is depicted for operational modes in the specified environments, and each function is described from both inter- and intrasystem viewpoints. To distinguish between functions and subfunctions, functions are indentured and identified from top down so that subfunctions are recognized as part of larger functions. Subfunctions are derived in an iterative process concurrent with the establishment of performance requirements, the development of effectiveness simulations, and the synthesis of progressively lower-level subsystem elements.

The fourth step of the systems engineering process is the *requirements allocation*. Each function, or subfunction, is allocated a set of requirements from top down. These requirements represent a set of minimum acceptable levels of performance for the accomplishment of the functions. All of the requirements are defined in sufficient detail for use as criteria for hardware design, equipment operation, personnel skills, facility operation, computer programming, display of data, and logistic support. The performances are derived in an iterative process with the system requirements, development of functions, synthesis and evaluation performed through trade studies and application of effectiveness models. In the previous high-performance aircraft example, the requirements allocation analysis would decide whether certain functions should be accomplished manually, automatically, or semiautomatically. Then when human operator control is required, there would be an allocation of the duties among the crew members.

The fifth step consists of *considering alternative solutions* by means of *system trade-off studies*. An experienced and imaginative systems engineer can synthesize several solutions to a problem. The more solutions he considers initially, the greater the probability of success for the final system. After weighing each solution carefully, only a handful of the most promising should be given further consideration. Practically speaking, the number of promising solutions should not exceed three, and for each of these, the systems engineer must compare the functional and the resulting hardware requirements of the system.

Effectiveness analysis simulations are performed in the following step. The simulation models allow the input parameters to be varied individually so that their relative effect on total system performance, reliability, maintainability, schedule, and cost can be determined. Having once established these simulation models, they should be maintained, updated, and modified during the progress of the program as required.

The *choice of the most optimum system* design is the seventh step. This is accomplished based on the results of the trade-off studies and effectiveness analysis simulations. The characteristics and data used in making this decision should be as accurate as humanly possible. If the information is unreliable, a decision cannot be made and it will be impossible to proceed further. The characteristics and data must be accurate enough to permit a clear differentiation among the merits of the various solutions proposed.

The *synthesis of the system* is the next step. This consists of the complete theoretical and physical design of the system.

The ninth step consists of *updating the equipment characteristics and data* for the system designed in the previous step. This is a continuing effort in the evolution of the overall system. As the program proceeds, less emphasis is placed on mathematical representations, computing, and simulation, while more emphasis is placed on the actual physical hardware being developed.

Testing of the system is the tenth step in the process. Determining the characteristics of the resulting physical system developed is a very important step. Good testing techniques and facilities are important attributes for a successful organization.

The final step consists of *correlating the test data of the system with the system objectives and requirements, and refining the design*. This step is very important and actually requires a reevaluation of the overall system interrelationships as illustrated in Figure 1.3. Certain goals may not be obtainable and a reexamination of previous assumptions and compromises may be necessary to meet the overall system requirements.

Summarizing, the general procedure involved in engineering an overall large complex system consists of:

1. Conceptual analyses.
2. Mission and requirements analyses.
3. Functional analyses.
4. Requirements allocation.
5. Trade-off studies.
6. Effectiveness analysis simulations.
7. Choice of the most optimum system design.
8. Synthesis of the system.
9. Updating the equipment characteristics and data.
10. Testing of the system.
11. Correlating the test data of the system with the system objectives and requirements, and refining the design.

This logical unified systems engineering procedure indicated in Figure 1.4 is really a feedback process. This characteristic is caused by the iterative characteristics of the solution and the fact that many aspects of systems engineering consist of verifying previous steps.

References

1. Eckman, D.P., ed. *Systems: Research and Design*. New York: John Wiley & Sons, Inc., 1961.
2. Peschon, J., ed. *Disciplines and Techniques of Systems Control*. New York: Blaisdell Publishing Co., 1965.
3. Licklider, J.C.R., "Man-Computer Symbiosis." *IRE Transactions on Human Factors in Electronics*, March 1960.
4. Shinners, Stanley M. *Control System Design*. New York: John Wiley & Sons, Inc., 1964.
5. Chestnut, Harold. *Systems Engineering Tools*. New York: John Wiley & Sons, Inc., 1965.
6. Morton J.A., "Integration of Systems Engineering with Component Development." *Electrical Manufacturing* 64, (August 1959) pp. 85-92.
7. Military Standard, *System Engineering Management*, MIL-STD-499 (USAF), 17 July 1969.
8. Mayr, O. "Adam Smith and the Concept of the Feedback System." *Technology and Culture* 12:1 (January 1971) pp. 1-22.

2 Performance

Introduction

The objective of this chapter is to indicate how the performance of a system may be determined. Concepts will be presented in this discussion regarding systems behavior as a function of its various parameters. This will provide the basis for comparing alternative solutions. Through this process, successive cycles of iteration will result in a progressively improved system design.

The modern systems engineer is concerned with large and complex systems. They are characterized by having a large number of inputs, outputs, and functions to be performed. In general, a change in one of these input variables produces a change in many of the system's output variables. Therefore, the systems engineer must concern himself with large systems having interacting inputs and outputs.

Modern systems incorporate a variety of physical and human elements in a configuration designed to achieve the required characteristics. A very important part of the systems engineering effort is to identify the appropriate deterministic or probabilistic models for the various components of the system. The mathematical models must adequately describe the dynamic transfer characteristics of the system's individual components to enable the systems engineer to determine system performance. Identification is an extremely important problem facing the systems engineer.

Another important characteristic of large and complex systems, fundamental in determining performance, is that many of the inputs to the system are statistical. For example, radar returns to an instrumentation tracking radar, demand for passenger space in an airline reservation system, or arrival and departure of airplanes in an air traffic command and control system, are all statistical. Therefore, performance of these systems can be determined only by utilizing statistical techniques, and the measure of performance will involve some probability distribution.

Large and complex systems are the objects of great interest today, and it is very important to determine their performance as precisely as possible. The basic elements concerned with determining system performance will be discussed. Several examples are provided to illustrate the application of these techniques.

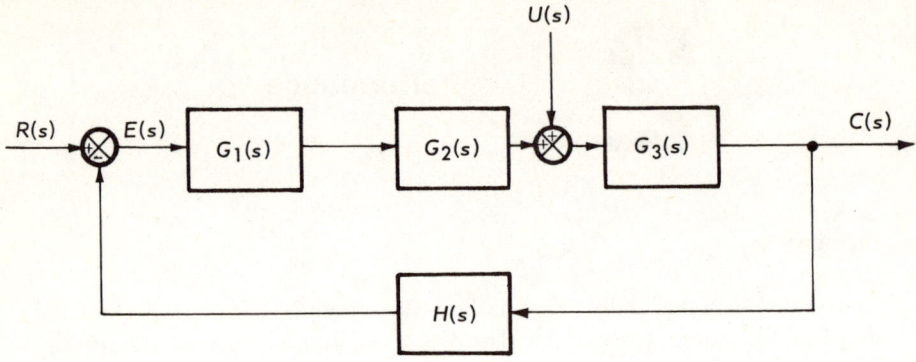

Figure 2.1. A Feedback Control System with an Undesired External Disturbance Input, $U(s)$.

Sources of System Error

There are many sources of error in large complex systems. In general, system error is caused by dynamic lags, disturbances, manufacturing tolerances, aging, and variations caused by environmental changes. If we consider the simple control system illustrated in Figure 2.1, we know that a reference input, $R(s)$, will result in an actuating signal, $E(s)$. We also know that a disturbance, $U(s)$, will also contribute to the error. The characteristics of these errors are discussed on pp. 30-37. In both open-loop and closed-loop systems, disturbances in the form of noise affect the system error. Manufacturing tolerances contribute to the error of all forms of systems. Practical considerations such as technological developments, cost, and schedule, require the engineer to specify components and equipment with finite tolerances. Aging and environmental changes (such as temperature and humidity) and line voltage and line frequency variations also contribute to the overall system error.

To illustrate this discussion, let us consider errors associated with specific components common to most control systems. For example, the use of integrating operational amplifiers is common to these systems. These elements are characterized by drift and offset (bias) errors. Potentiometer errors are characterized by their linearity, resolution, and noise. Synchro errors are characterized by their resolution and null noise voltage. Tachometer errors are characterized by their sensitivity changes as a function of temperature and quadrature voltage.

The use of gyros, as for stable platforms, presents several sources of error. It has a random drift rate in addition to a bias error. Furthermore, it produces errors caused by friction, hysteresis, imperfect temperature regulation, signal generator misalignments, and cross-coupling effects.

The use of accelerometers, as in inertial navigation systems, presents several sources of error. The accelerometer produces errors due to bias errors and random drift. It also causes errors due to the effects of friction, hysteresis, cross-coupling effects, and imperfect temperature regulation.

Stable platforms, used in airborne and ship-based applications, produce system errors caused by misalignments between the axis, reference directions, and misalignments between the components, such as gyros. In addition, any weight imbalances in the platform will also result in system errors.

How do we specify system error? Consider the following two specific systems: an inertial navigation system and a tracking radar system.

For the case of an inertial navigation system associated with a manned aircraft, system error is usually defined as the rate of increase of position error, such as one nautical mile per hour. Inertial navigation systems for ballistic missiles usually are characterized in terms of the position error at the impact point. Alternative specifications sometimes include the error of the velocity vector at cutoff. In the case of orbital satellite applications, the system error is usually defined in terms of the velocity error at cutoff. For the case of interplanetary space vehicle applications, system error refers to the miss distance of the space vehicle from the target planet.

Next, consider the case of a tracking radar system. For this system, error is usually defined in terms of its dynamic, random, and system-associated error components. Dynamic errors result from control system lags generated by reference input velocities, accelerations, and rate of change of acceleration. These errors are characterized by having a variable and constant component. Random errors are associated with such factors as tracking radar jitter, radar mount, and dish flexure caused by wind loads, noise on data pickoffs, etc. They are characterized by random fluctuations about zero mean.[1] System-associated errors are primarily caused by system misalignments or by other noncontrollable causes such as the refraction of a radar beam.

An error budget is usually associated with each radar which determines its capability as a function of several parameters and conditions. In complex systems, such as a tracking radar, overall system-associated error cannot be specified by only one number. Instead, an error budget table is usually established from which system-associated error can be determined as a function of several characteristics and conditions.

The Identification Problem

Before system performance can be determined, the dynamic equations of the process must be known. These mathematical models must describe the dynamic transfer characteristics for the individual elements of the system.

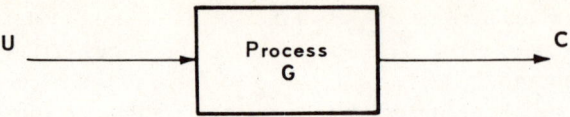

Figure 2.2. Identification Problem of an Open-Loop System

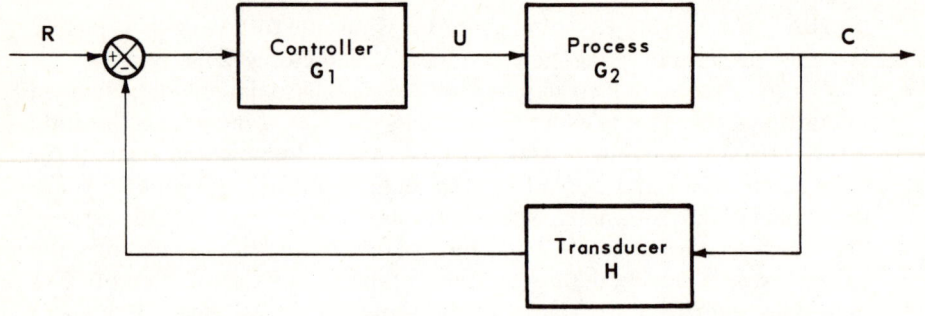

Figure 2.3. Identification Problem in a Closed-Loop System

Therefore, the identification problem first involves a determination of the appropriate form of each model. Next, these models must be evaluated from analysis of the physical laws or experimental measurements of the element's characteristics. Third, the initial model must be modified to include any statistical variations due to manufacturing tolerances, aging, and environmental conditions. The element can be considered adequately identified only when its mathematical characteristics are complete enough to permit the analysis of system performance.[2]

To identify a processes-dynamical behavior for performing an analysis of anticipated performance, extensive measurements are made *a priori* to system development. Consider the open-loop process illustrated in Figure 2.2. The objective of the systems engineer is to evaluate the transfer characteristics from the specific inputs to the outputs of interest. The set of transfer characteristics, represented by matrix G, can be obtained from the measured values of the multivariable inputs U and multivariable outputs C. Figure 2.3 illustrates the problem of identification in a closed-loop system. To design this system for proper performance, the exact characteristics of the process G_2 must be known. This matrix of dynamic relationships among each of the actuating signals and outputs can be determined from measurements of U and C.

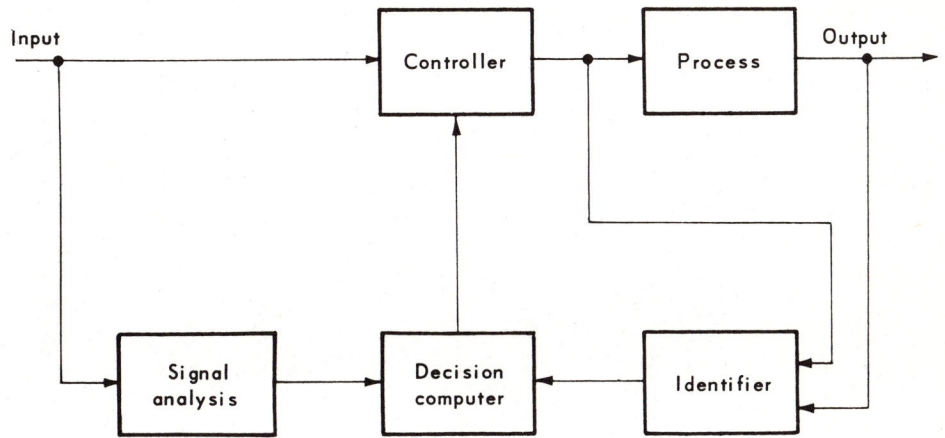

Figure 2.4. An Adaptive System

So far we have talked about identification in an *a priori* sense. That is, we are concerned in the identification of the process prior to system operation. There is a large class of systems, designated as "adaptive systems" which is concerned with identification during operation.[2] An adaptive system is one that automatically achieves a desired response in the presence of extensive changes in the process parameters. Such systems are usually characterized by devices which automatically measure the dynamics of the process, and other devices which automatically adjust the characteristics of the controller based on a comparison of these measurements with some optimum figure of merit.[3] Adaptive systems, which can be open-loop or closed-loop, have stimulated work in the area of identification.

Figure 2.4 illustrates an adaptive system. An "identifier" block measures the transfer characteristics of the process continuously in real time during operation. This information, together with an analysis of the input signal, is fed to the decision computer which then determines the desirable characteristics for the controller based on a performance criterion which must be satisfied.

If the process can be represented as a linear, time-invariant device, then the problem of identification can be readily solved utilizing well-known techniques.[4] These methods can evaluate the frequency response, impulsive response, and resulting differential equation. However, if the device is nonlinear and/or time-variable, then the problem of identification is very difficult. Reasonable approximations can be made by utilizing *a priori* information regarding the form of the transfer characteristics of the device and considering only significant terms.

Predictable and Unpredictable Errors

System errors are usually classified as being either predictable or unpredictable. Predictable system errors are those errors which can be foreseen based on the mathematical description of processes dynamics. Unpredictable errors are generally random and require statistical techniques to describe their characteristics.

Predictable system errors are known for all input conditions and values of time. These errors can be compensated for and subtracted out of the system with a data processor. This is done only if errors exceed some threshold limit. Usually the system is designed so that the predictable errors lie within tolerable limits. In large complex systems, however, where a data processing system is usually available, it is a simple matter to correct and compensate for these errors.

If prior measurements and known characteristics do not permit the accurate prediction of future errors, the error is characterized as unpredictable. These errors must be treated using statistical methods to obtain a mathematical description of them. Based upon measured data, the random errors associated with most large complex systems, such as navigation and tracking radar systems, are usually characterized by having a Gaussian probability density. This simple statistical distribution enables use of well-developed mathematical tools for applying statistical techniques to the study of such resulting system errors.

Predictable Errors

Many components of a system's error are predictable. The lag following errors, normally associated with a feedback control system, fall into this category. Other predictable control system errors depend on the particular system under consideration. For example, errors caused by the external disturbance signal $U(s)$ entering the feedback control system, illustrated in Figure 2.1, are predictable. In general, any error which can be determined for all input conditions and values of time is classified as a predictable error.

To illustrate the techniques involved in treating predictable errors, let us reconsider the feedback system illustrated in Figure 2.1. Theoretically, it is desirable for this control system to have the capability of responding to changes in position, velocity, acceleration, and to higher-order derivatives with zero error. The closed-loop system transfer function, $T(s)$, is defined by Equation (2.1), where $R(s)$ and $C(s)$ represent the input and output, respectively.

$$T(s) = \frac{C(s)}{R(s)}. \tag{2.1}$$

By definition, the error $E(s)$ is given by

$$E(s) = R(s) - C(s). \tag{2.2}$$

For a unity feedback system $[H(s) = 1]$, $E(s)$ can be identified with the actuating signal and can be described by an infinite series, such as Equation (2.3):

$$E(s) = R(s)\left(\sum_{n=0}^{\infty} \frac{1}{K_n} s^n\right). \tag{2.3}$$

The K_ns are commonly referred to as follows:

$$K_0 = 1 + K_p = 1 + \text{position constant}, \tag{2.4}$$

$$K_1 = K_v = \text{velocity constant}, \tag{2.5}$$

$$K_2 = K_a = \text{acceleration constant}. \tag{2.6}$$

Higher order terms of the series can be ignored if the series converges rapidly for a stated input.

To synthesize a system configuration, it is desirable to express the error coefficients K_n in terms of the parameters of the closed-loop system transfer function, $T(s)$.[4,5] Substituting Equation (2.3) into Equation (2.2), we obtain

$$C(s) = R(s)\left(1 - \sum_{n=0}^{\infty} \frac{1}{K_n} s^n\right). \tag{2.7}$$

From the definition given by Equation (2.1), Equation (2.7) can be expressed as follows:

$$T(s) = \left(1 - \sum_{n=0}^{\infty} \frac{1}{K_n} s^n\right). \tag{2.8}$$

Consider $T(s)$ as expressed by the ratio of two factorable polynomials in s as in Equation (2.9)

$$T(s) = K \frac{\left[\prod_{u=1}^{A} (s + Z_u)\right]\left[\prod_{m=1}^{B} (P_m)\right]}{\left[\prod_{m=1}^{B} (s + P_m)\right]\left[\prod_{u=1}^{A} (Z_u)\right]}, \tag{2.9}$$

where A is the order of the numerator, and B is the order of the denominator. Equating Equations (2.8) and (2.9) it is possible to express K_n in terms of Z_u and P_m which represent the gain, zeros and poles of the closed-loop system transfer function, respectively.

$$\sum_{n=0}^{\infty} \frac{1}{K_n} s^n = 1 - K \frac{\left[\prod_{u=1}^{A} (s + Z_u)\right]\left[\prod_{m=1}^{B} (P_m)\right]}{\left[\prod_{m=1}^{B} (s + P_m)\right]\left[\prod_{u=1}^{A} (Z_u)\right]}. \quad (2.10)$$

Taking the limit of both sides of Equation (2.8) as s approaches zero, the following is obtained:

$$\frac{1}{K_0} = 1 - K \quad (2.11)$$

or

$$K_0 = (1 + K)^{-1} \stackrel{\Delta}{=} 1 + K_p. \quad (2.12)$$

This is the classical result for a type 1 system ($K = 1$ and $K_0 = \infty$). The first derivative of Equation (2.10) reduces to the following:

$$\frac{1}{K_1} = \frac{d}{ds}\left\{-K \frac{\prod_{m=1}^{B}(P_m)}{\prod_{u=1}^{A}(Z_u)}\left[\frac{\prod_{u=1}^{A}(s+Z_u)}{\prod_{m=1}^{B}(s+P_m)}\right]_{s=0}\right\}. \quad (2.13)$$

For type 1 or higher systems, $T(s) = 1$ when $s = 0$ and

$$\left[\frac{d}{ds} T(s)\right]_{s=0} = \frac{\left[\frac{d}{ds} T(s)\right]_{s=0}}{[T(s)]_{s=0}} = \left[\frac{d}{ds} \ln T(s)\right]_{s=0}$$

Therefore, using this relationship in Equation (2.3), the following can be derived:[5]

$$\frac{1}{K_1} = \frac{1}{K_v} \stackrel{\Delta}{=} \sum_{m=1}^{B} \frac{1}{P_m} - \sum_{u=1}^{A} \frac{1}{Z_u} \quad (2.14)$$

In a similar manner, it can be shown that

$$-\frac{2}{K_2} = \frac{2}{K_a} \stackrel{\Delta}{=} \frac{1}{K_1^2} + \sum_{m=1}^{B} \frac{1}{P_m^2} - \sum_{u=1}^{A} \frac{1}{Z_u^2} \quad (2.15)$$

For the great majority of system applications, the systems engineer is not concerned with the input's rate of change of acceleration, or any higher order derivative inputs. Therefore, let us focus attention in this section on the resulting errors caused by system inputs corresponding to position, velocity, and acceleration.

Defining the well-known position, K_p, velocity, K_v, and acceleration, K_a, constants by[4]

$$K_p = \lim_{s \to 0} G(s)H(s) \qquad (2.16)$$

$$K_v = \lim_{s \to 0} sG(s)H(s) \qquad (2.17)$$

$$K_a = \lim_{s \to 0} s^2 G(s)H(s) \qquad (2.18)$$

where $G(s)$ represents the combined transfer function of $G_1(s)$, $G_2(s)$, and $G_3(s)$ for the system shown in Figure 2.1.

The steady-state errors, $e(t)_{ss}$ to unit inputs and unity feedback are given by

$$e(t)_{ss} \bigg]_{\substack{\text{Unit} \\ \text{Step} \\ \text{Input}}} = \frac{1}{1 + K_p} \quad \text{for a type 0 system} \qquad (2.19)$$

$$e(t)_{ss} \bigg]_{\substack{\text{Unit} \\ \text{Ramp} \\ \text{Input}}} = \frac{1}{K_v} \quad \text{for a type 1 system} \qquad (2.20)$$

$$e(t)_{ss} \bigg]_{\substack{\text{Unit} \\ \text{Parabolic} \\ \text{Input}}} = \frac{2}{K_a} \quad \text{for a type 2 system} \qquad (2.21)$$

A summary of these and other constants for types 0, 1, 2, and 3 control systems is listed in Table 2.1 If the inputs are other than unit quantities, the steady-state errors shown in Equations (2.19) through (2.21) are proportionally increased. For example, if the input to a type 1 system is a ramp whose value is B position inputs (feet, yards, etc.)/second, then the steady-state error as given by Equation (2.20) would be modified to read $e(t) = B/(1 + K_v)$.

Another common predictable system error is caused by undesired external disturbances entering a feedback control system. The relationship between the resulting system error $E_U(s)$, and the undesired disturbance input, $U(s)$, with $R(s) = 0$ for the system shown in Figure 2.1, is given by

$$\frac{E_U(s)}{U(s)} = \frac{-G_3(s)H(s)}{1 + G_1(s)G_2(s)G_3(s)H(s)}. \qquad (2.22)$$

Assuming that $U(s)$ is measurable, the resulting system error can be determined for various types of disturbances.

An important feature of calculable errors is that they can be compen-

Table 2.1
Summary of Steady-State Error Values for Various Types of Inputs

	Type of Input		
System Type	Unit Step	Unit Ramp	Unit Parabolic
0	K_p	0	0
1	∞	K_v	0
2	∞	∞	K_a
3	∞	∞	∞

sated for and subtracted out of the system with a data processor if they exceed some threshold level. The resulting errors of the system shown in Figure 2.1, due to the unwanted external disturbance that occurs at a point which is inaccessable electrically, can be compensated for as shown in Figure 2.5. It is assumed that $U(s)$, which is measurable, enters at a point which represents torque and an electrical compensation signal cannot enter this point directly. For this case, the resulting effect of the compensation signal at the error point differs from the effect of $U(s)$ by the factor $1/G_2(s)$. The relationship between the external compensation signal, $U_c(s)$, and the system actuating signal $E_{U_c}(s)$ is given by

$$\frac{E_{U_c}(s)}{U_c(s)} = \frac{\cancel{G_2(s)}G_3(s)H(s)}{1 + G_1(s)G_2(s)G_3(s)H(s)} \cdot \frac{1}{\cancel{G_2(s)}}. \quad (2.23)$$

By adding Equations (2.22) and (2.23), we find that the net effect on system error due to $U(s)$ and $U_c(s)$ is cancelled:

$$E_U(s) + E_{U_c}(s) = 0. \quad (2.24)$$

Therefore, the calculable error due to the unwanted external disturbance has been compensated for by means of the data processor in a closed-loop manner. The correction is usually never perfect, primarily due to inaccuracies in the measurements of $U(s)$, the identification problem associated with determining $G_2(s)$, manufacturing tolerances, aging, and variations caused by environmental conditions.

A data processor can also be used to compensate for the error in an open-loop manner as shown in Figure 2.6. The primary disadvantage of open-loop processing is that the technique does not correct the closed-loop errors. This may not be acceptable in many cases. For example, in a tracking radar problem, open-loop compensation will not prevent the loss of target tracking resulting from dynamic lag errors which exceed the allowable tracking error limits. For these cases, nonlinear and adaptive tracking loops are required to position the radar beam on the target.[5]

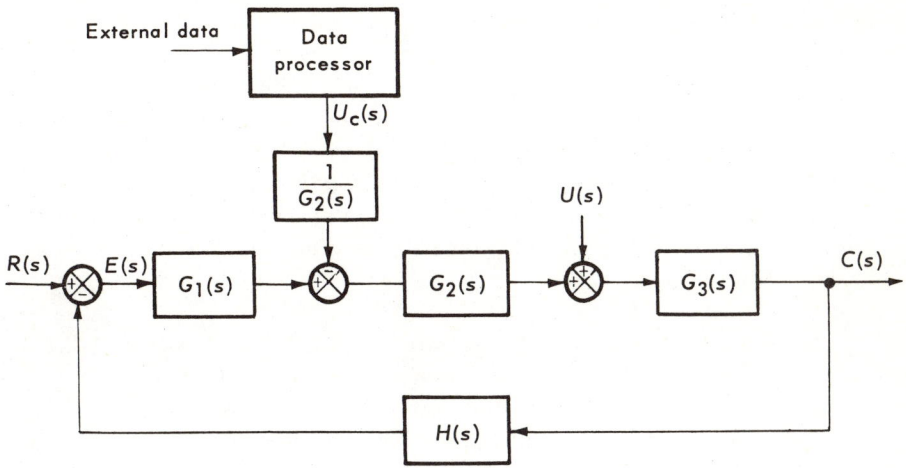

Figure 2.5. Correcting the System Error Due to an External Disturbance in a Closed-Loop Manner

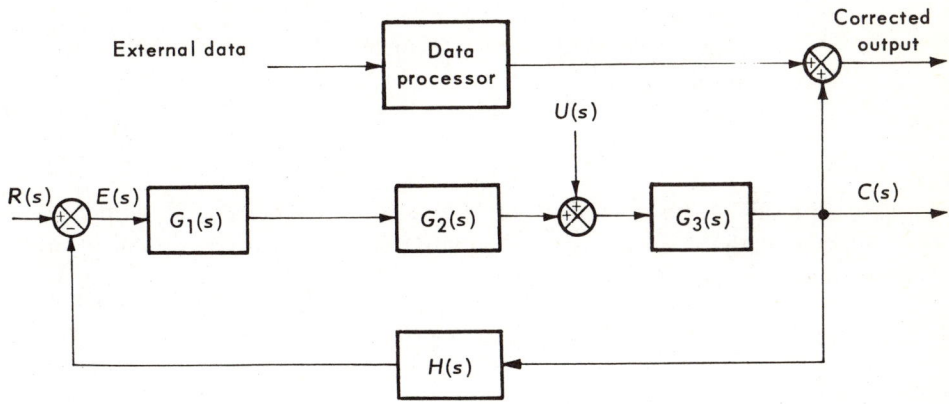

Figure 2.6. Correcting the System Output in an Open-Loop Manner

The amount of computer capacity required for this correction is usually very small for most common applications. However, in some cases the computer implementation can become complex. For these cases, it is recommended that the characteristics be approximated by a simpler equation in order to ease the data processing problem. This is usually an acceptable compromise.

Characteristics of Unpredictable Errors

Let us review some of the basic tools of probability theory in order to understand the methods to be presented for treating unpredictable system errors. Consider the *probability density function, p(x)*, of a continuous random variable. Let us define the accumulated probability density distribution that the random variable, X, takes on a value in the interval $x \leq X \leq x + dx$ as $p(x)\,dx$. The probability density function satisfies the following relationship:

$$\int_{-\infty}^{\infty} p(x)\,dx = 1 \tag{2.25}$$

where $p(x) \geq 0$. The mean value is a convenient mathematical measure of the value about which the random variable X is concentrated. It is defined by the expression

$$\bar{x} = \int_{-\infty}^{\infty} x p(x)\,dx. \tag{2.26}$$

The *dispersion of the probability* density distribution is a measure of the scattering of the values about the mean. It is usually expressed in terms of the moments of $p(x)$ about the arithmetic mean $\bar{x}$. Its defining relationship is given by the following:

$$\bar{x}^n = \int_{-\infty}^{\infty} (x - \bar{x})^n\, p(x)\,dx. \tag{2.27}$$

In all cases, $\bar{x}^0$ equals unity and $\bar{x}^1$ equals zero. The second moment is defined as the variance, σ^2. The square root of the *variance* is called the *standard deviation*, σ.

The *Gaussian probability density* distribution is a very important probability density distribution, since it occurs in many practical systems. It is defined by the expression

$$p(x) = \frac{1}{\sqrt{2\pi}\sigma} \exp\left[-\frac{(x - \bar{x})^2}{2\sigma^2}\right] \tag{2.28}$$

where $\bar{x}$ is the mean value of the probability density distribution and σ^2 is the variance.

Consider the bell-shaped Gaussian probability density distribution curve shown in Figure 2.7. The mean value of the probability distribution, denoted by $\bar{x}$, is obtained by computing the average value of the random variable x. In practical systems, such as inertial navigation systems and tracking radar systems, the mean value is generally referred to as the bias value and is usually predictable with a fair degree of accuracy. In practice,

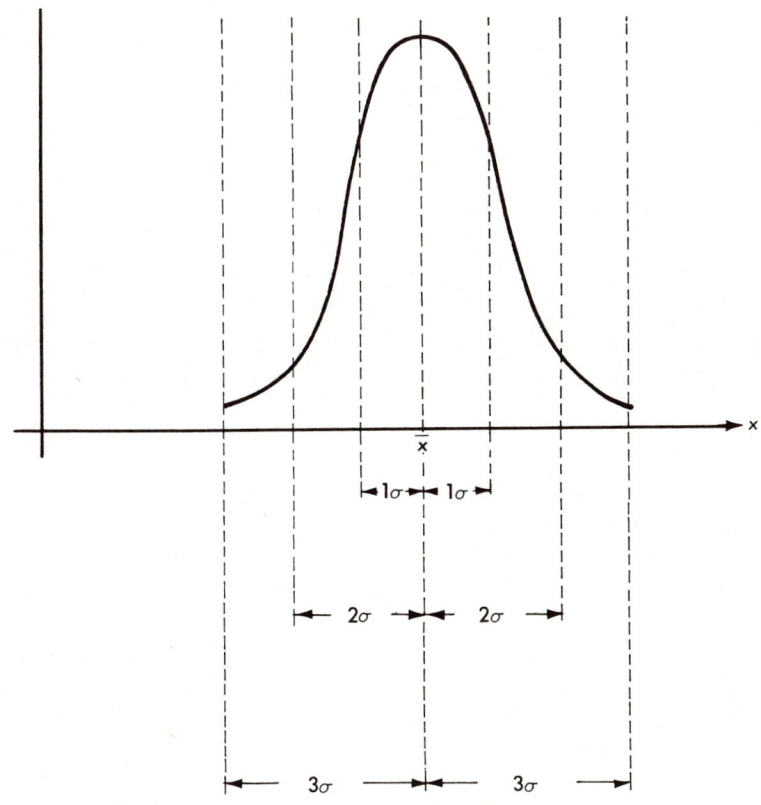

Figure 2.7. A Gaussian Probability Density Distribution

therefore, this bias error component usually can be compensated for and is assumed to be essentially zero.

The significance of the 1-sigma value is that it represents 68.26 percent of the area under the bell-shaped curve. This area is equally spaced for positive and negative deviations from the mean value, $\bar{x}$. Therefore, it represents the variation from average in the positive and negative directions. The 2-sigma value represents 95.46 percent of the area under the Gaussian curve. Furthermore, the 3-sigma value represents 99.73 percent of the area under this curve.

Therefore, when a Gaussian error is denoted by its 1-sigma value, it signifies that the random function x will lie within the ± 1-sigma band 68.26 percent of the time. Similarly, when the Gaussian error is denoted by the 2-sigma value, it signifies that the random function x will lie within the ±

2-sigma band 95.46 percent of the time. In the case of a Gaussian error denoted by its 3-sigma value, it signifies that the random function x will lie within the $\pm$ 3-sigma band 99.73 percent of the time. In other words, the random function will not be within the 3-sigma value band only 27 times out of 10,000 measurements.

Other useful designations of the random Gaussian curve are the *mean absolute error* (m.a.e.) and the *probable error* (p.e.). The m.a.e. is defined as

$$\text{m.a.e.} = 0.7979\sigma. \qquad (2.29)$$

Its significance is that it is the mean of all the deviations from the mean without regard to sign. It can be computed from:

$$\text{m.a.e.} = \int_{-\infty}^{\infty} |x - \bar{x}| p(x)\, dx$$

$$= \int_{-\infty}^{\bar{x}} (\bar{x} - x) p(x)\, dx + \int_{\bar{x}}^{\infty} (x - \bar{x}) p(x)\, dx. \qquad (2.30)$$

The p.e. is defined as

$$\text{p.e.} = 0.6745\sigma. \qquad (2.31)$$

Its significance is that fifty percent of the area under the $p(x)$ curve lies above and below this value. Mathematically,

$$\text{p.e.} = \int_{\bar{x}-0.6745\sigma}^{\bar{x}+0.6745\sigma} p(x)\, dx = 0.5. \qquad (2.32)$$

Although the Gaussian probability density distribution occurs most frequently in practice, several other probability density distributions are found. These include the Poisson, binomial, and exponential probability density distributions. We define them in the remainder of this section and illustrate some practical applications.[9]

The *Poisson distribution* is another useful model in practice. The definition of this distribution is given by

$$P(x) = \frac{\bar{x}^{2x}}{x!} e^{-\bar{x}^2}. \qquad (2.33)$$

It is used in practice for representing a random independent event that occurs over equal periods of time at a constant average rate. For example, it has been used to represent the random distribution of the number of calls per hour received during a particular time of day at a large department store, the arrival of aircraft at an airport, the number of flaws in similar pieces of metals, and other independent variables which can be represented over a constant area or volume.

The *binomial probability distribution* is useful for determining the probability distribution of a function that has the characteristics of being successful X times in n independent experiments, where the probability of success per trial is a constant value p. The mathematical expression for the binomial probability distribution is given by

$$P(x) = \frac{n!}{x!(n-x)!} p^x (1-p)^{n-x} \qquad (2.34)$$

where $0 \le p \le 1$. The binomial distribution has been used in practice to determine the probability of success for missile firings. For example, if 20 independent missile firings are planned, each having a probability of success of 0.98, the binomial distribution can be used to determine the probability that exactly 18 of the experiments will be successful.

The *exponential density distribution* is very useful in practice for determining the time between occurrences of independent random events. The definition of this function is given by

$$P(x) = re^{-rx} \qquad (2.35)$$

where r is the distribution parameter and $x \ge 0$. The exponential distribution is utilized quite frequently in practice to determine the failure of an element or complex system. It will be discussed in more detail in Chapter 3 when we discuss reliability.

Combining Errors

Large complex systems have several sources of error. As discussed previously, system errors can generally be classified as being composed of dynamic, random, and systems-associated errors. This section indicates how the various system errors can be combined.

Dynamic Errors

As an example, consider a system whose change, ΔP, in systems dynamic performance $P(p_1, \ldots, p_n)$ is a linear function of the sum of changes to several individual parameter error changes in dynamic performance Δp_i ($i = 1, \ldots, n$). Assuming that n is the total number of individual sources of error, then the change in system performance is given by the following linear series.

$$\Delta P = \Delta P_1 + \Delta P_2 + \Delta P_3 + \ldots + \Delta P_n \qquad (2.36)$$

The resulting system dynamic error, E_D, is given by

$$E_D = E_1 + E_2 + E_3 + \ldots + E_n \tag{2.37}$$

where $E_i (i = 1, \ldots, n)$ represents the resulting error caused by the changes of the individual points of the parameter p_i.

Random Errors

A large complex system usually has several sources of unpredictable random errors which are independent of each other. Assuming that these sources are in cascaded elements, how does the control systems engineer combine these errors? Straight addition of the statistical values would result in an extremely pessimistic description of the probable error. To overcome this problem, statistical errors that are independent of each other are combined by the root-mean-square (RMS) technique. Assuming that $\sigma_1, \sigma_2, \sigma_3, \ldots, \sigma_n$ are the variances of independent random errors of zero mean which combine to cause an overall system error, then the total error is denoted as the root-mean-square error variance σ_{RMS}, which is given by

$$\sigma_{\text{RMS}} = \sqrt{\sigma_1^2 + \sigma_2^2 + \sigma_3^2 + \ldots + \sigma_n^2} \tag{2.38}$$

where the individual variances are the σ Gaussian density values.

The root-mean-square error is sufficient to define the random error for one-dimensional systems. In the case of two-dimensional problems, additional statistical characteristics must be defined. For example, in an inertial navigation system, a root-mean-square value can be obtained for the accumulation of "along-course" (range) errors, $1\sigma_{ac}$, and an additional root-mean-square value can be obtained for the "cross-course" (lateral) errors, $1\sigma_{cc}$. Considering each of these independent statistical errors separately, the Gaussian distribution can be used to describe the characteristics of the errors. However, to combine σ_{ac} and σ_{cc}, the systems engineer must resort to a probability ellipse to describe the characteristics of the unpredictable random errors. The probability ellipse reduces to a probability circle if $\sigma_{ac} = \sigma_{cc}$.

The problem becomes more complicated for three-dimensional problems. In the case of a tracking radar, for example, the systems engineer is concerned with unpredictable errors in azimuth, elevation, and range. The problem can be approached from two directions. First, the problem can be simplified by considering the unpredictable random error to be characterized by an angular and range error. For the angular error, the control systems engineer would combine the root-mean-square errors in azimuth and elevation to obtain a probability ellipse. The range error can remain as a RMS value. A second, more complicated approach is to consider the total

Table 2.2
Relationship Between CPE and Orthogonal Normal Distributions[7]

σ_{cc}/σ_{ac}	CPE/σ_{ac}	σ_{cc}/σ_{ac}	CPE/σ_{ac}
0.05	0.683	0.55	0.904
0.10	0.688	0.60	0.935
0.15	0.698	0.65	0.966
0.20	0.713	0.70	0.996
0.25	0.732	0.75	1.026
0.30	0.753	0.80	1.056
0.35	0.782	0.85	1.087
0.40	0.812	0.90	1.118
0.45	0.843	0.95	1.149
0.50	0.872	1.00	1.180

volumetric error caused by the azimuth, elevation, and range RMS components. An ellipsoid then results which describes the unpredictable random errors in three dimensions.

Returning to the probability ellipse case for two dimensions, consider the probability, P, that some random point (x, y) will fall within the ellipse. Mathematically, P is given by[6]

$$P = 1 - e^{-0.5C^2}. \qquad (2.39)$$

When $C = 1.774$, the probability P is 0.5, and the corresponding ellipse is called the 50 percent probability ellipse. In the inertial navigation problem, when $\sigma_{ac} = \sigma_{cc}$, the 50 percent probability circle is the circle of radius 1.774σ, where $C = 1.1774$ and $P = 0.5$. The radius of this circle is called the circular probable error (C.P.E.). It is related to σ by

$$\text{C.P.E.} = 1.1774\sigma. \qquad (2.40)$$

As an example of applying the CPE concept, consider the case of an inertial navigation system where $\sigma_{ac} = \sigma_{cc} = 1.0$ n.mi. For this case, the CPE can be obtained from Equation (2.40) and equals 1.1774 n.mi. For the cases where $\sigma_{ac} \neq \sigma_{cc}$, the computation becomes more difficult. For these cases, the ratios shown in Table 2.2 are very useful for computation of the CPE.[7] By convention, it is common to refer to the RMS error in two dimensions as the CPE, even for cases where $\sigma_{ac} \neq \sigma_{cc}$ and the probability distribution is elliptical rather than circular. To illustrate how to use Table 2.2, consider the example where $\sigma_{ac} = 1$ n.mi. and $\sigma_{cc} = 0.4$ n.mi. Their ratio is given by

$$\frac{\sigma_{cc}}{\sigma_{ac}} = 0.4.$$

From Table 2.2, we find that

$$\frac{\text{CPE}}{\sigma_{ac}} = 0.812.$$

Since $\sigma_{ac} = 1$ n.mi., the value of CPE is given by 0.812 n.mi.

System Associated Errors

Systematic errors are basically caused by changes due to manufacturing tolerances, aging, and variations caused by environmental conditions. For example, systematic errors are caused by the change in linearity when certain limits are exceeded, changes in dimensions due to temperature, alignment changes caused by wear, reference changes due to environmental conditions, and errors in large structures which are caused by changes in the gravitational forces due to weight unbalances.

The systematic errors can be combined to provide a resultant systematic component of error for each of the conditions of interest. For example, assuming that there are n sources of systematic errors, $E_{s_1}, E_{s_2}, \ldots, E_{s_n}$, then the total resultant systematic error, E_{s_T}, is given by

$$E_{s_T} = E_{s_1} + E_{s_2} + \ldots + E_{s_n}. \tag{2.41}$$

The proper sign must be included with each of these errors.

Resultant System Error

The resultant total system error, E_T, caused by the simultaneous presence of dynamic, random, and systematic errors, can be obtained in several manners. The most accurate technique is to consider the separate components for each of these errors. A frequently used approximation, however, is to take the square root of the sum of the squares of these three components, as follows:

$$E_T = \sqrt{E_D^2 + \sigma_{\text{RMS}}^2 + E_{s_T}^2} \tag{2.42}$$

where E_D, σ_{RMS}, and E_{s_T} are defined by Equations (2.37), (2.38), and (2.41), respectively.

In many situations where multiple probability distributions exist, the simple approximation given by Equation (2.42) is not valid, and the resultant system error must be calculated in a more exact manner. The Monte Carlo technique,[8,9,11] discussed in Chapter 5, pp. 120-124, is a very useful technique for accomplishing this more exact determination.

The Error Budget

As we have discussed, there are several sources of errors in systems. For example, in an inertial navigation system, the systems engineer must contend with errors due to gyros, accelerometers, servos, the computer, the computer equations, initial alignments, geographical information, and gravity anomalies. In a tracking radar, the systems engineer must contend with errors due to target dynamics, nonlinear friction, receiver thermal noise jitter, antenna boresight shifts, wind gusts, mechanical misalignments, the data unit, uneven solar heating, weight unbalances, the data processor, and the data processor's equations. The systems engineer must consider the possible sources of dynamic, random, and systematic error during the preliminary engineering design phase. He must estimate the contribution of each source of error during the anticipated operating conditions, and then determine the overall system's predictable and unpredictable errors. These figures are then compared to the allowable overall system error to determine whether the estimated error contributions can be tolerated. If they are tolerable, the control systems engineer can then specify the allowable error for each contributing source. If they are not tolerable, he must go back and renegotiate his estimates, the system design, and/or the system's philosophy.

The term "error analysis" is used to denote the process of estimating, negotiating, and computing the total system error. After the various sources of error have been considered and are accepted as being tolerable, an error budget is specified. Each source of error in the system must adhere to the specifications of the error budget. In the event that a certain source of error exceed its specified error after being developed, the systems engineer must attempt to reduce other sources of error in the error budget or else increase the overall system error.

In general, the initially specified error budget is only a tentative one, since detailed component specifications are rarely available in practice. In addition, the state of the art for certain components as gyros, accelerometers, radar receivers, etc., is continually changing. After the survey of the available hardware is completed, a trade-off analysis of the initial error budget is made to determine a final error budget. This final step takes into account the present state of the art of the hardware, in addition to the detailed requirements of all components in the system, to adhere to the overall system's error requirements.

An Example—A Tracking Radar System

The three major sequential phases of operation for a tracking radar are denoted as designation, acquisition, and tracking.[1] During *designation*, the radar is programmed to the general location of the target within a certain accuracy which depends upon the available information. In the *acquisition phase*, the radar beam is scanned around this area in a preprogrammed scan pattern. The radar ceases scanning in its preprogrammed pattern and commences *automatic tracking* when the radar beam crosses a target. Each of these three phases of operation has certain accuracy requirements and an error budget associated with it. The most complex and important phase is automatic tracking and attention will be focused on this particular mode of operation in this section.

During automatic tracking, system error is defined as the angle between the apparent line of sight to the target and the center line of the tracking radar beam. The word "apparent" is used to describe these errors, since radar systems errors caused by atmospheric refraction and target scintillation are not included in this definition.[10] In addition, only the elevation axis of a two-axis elevation over azimuth tracking radar will be considered. Actually, this is the more complex axis from an error viewpoint. The tracking radar is assumed to have a 35-foot reflector and an inertia of 20,000 slug feet^2 in elevation. Furthermore, it is assumed that the radar does not use a radome and uses an encoder for output position transmission.

The sources of error in the elevation axis are caused by the following subsystems:

1. Antenna Subsystem
    a. Boresight
    b. Structural distortions due to uneven solar heating
    c. Structural distortions due to gravity
    d. Structural distortions due to wind
2. Mechanical Subsystem
    a. Leveling
    b. Bearing wobble
3. Servo Subsystem
    a. Dynamic lag
    b. Nonlinear friction
    c. Wind gusts
    d. Noise
4. Receiver Subsystem
    a. Thermal noise jitter
    b. Phase shift error
5. Data Unit Subsystem
    a. Encoder readout
    b. Noise

Each of these errors will be defined, classified as to their characteristics, and a typical value assigned to it. An error budget is then prepared to determine the overall tracking systems error.

Antenna Subsystem

The *boresight error* represents the angle between the *R-F* axis and the optical axis of the boresight telescope. This error also includes any error associated with the boresight telescope. The boresight error is classified as systematic. A typical value for this type of tracking radar, based on measurements of similar systems, is 0.002 degrees.

Structural distortion due to uneven solar heating results in significant system errors with tracking radars using large reflectors. Incident solar energy can cause heating at a thermal rate of approximately 350 BTU/(hr × ft^2). A few degrees of temperature differential can contribute a significant systems error for a precision tracking system with support points located several feet apart. To minimize this effect, a number of beams, highly reflective paint, fans, and insulation are utilized. Even with these precautions, a significant system error due to this source occurs. This error is characterized as being random. It is assumed that the 3σ error contribution is 0.005 degrees for this system, based on measurements of similar systems.[10]

Structural distortions due to gravity must also be accounted for in large tracking radar systems of this type. Normally, the axis of the antenna is boresighted in a nearly horizontal position. As the elevation angle is increased, structural deformations caused by the change in direction of the gravity forces cause a change in the elevation radar boresight. This error, which is characterized as being systematic and variable, is assumed to be 0.001 degrees, based on measurements of similar systems.[10] It can actually be measured and plotted as a function of elevation angle. Using a data processor, most of this error can be corrected.

Structural distortions due to wind result is significant errors in large tracking radar systems. For this reason, and because of the resulting servo error due to wind gusts, radomes are utilized when these errors become intolerable. In this example, however, it is assumed that a radome is not used. The wind causes various types of loading and various resulting errors as a function of the relative orientation of the antenna and the wind's direction. The effect of wind structural distortion in elevation increases as the zenith is approached and is minimal as the horizontal position is approached. The actual value of the error contributed is a function of the antenna's aerodynamic properties and the structural flexibility of the antenna. This error is characterized as being random and its 3σ value is assumed to be 0.01 degrees for this problem.

Mechanical Subsystem

Leveling errors in the elevation axis are caused by imperfect alignment of the radar's horizontal plane. This error is characterized as systematic. For this problem, the leveling error is assumed to be 0.01 degrees in the elevation axis, based on measurements in similar systems.[10]

Bearing wobble results in a random error. It is assumed for this problem that the 3σ bearing wobble error is 0.002 degrees, based on measurements of similar systems.

Servo Subsystem

The *dynamic lag* due to servo response has been discussed on pp. 24-29. For this problem, it is assumed that we have a type 1 system with a velocity constant of 1000. Furthermore, it is assumed that the maximum target angular velocity component that the radar has to track in elevation is 10 degrees per second and angular accelerations are negligible. This results in a maximum dynamic error of 0.01 degrees. This error is a function of range, target dynamics, and the relative geometry of the target to the radar. The error can be determined quite accurately and can be compensated for in a data processor, if desired.

Nonlinear friction, in addition to creating a stability problem due to its phase lag characteristics,[4] also acts as a torque disturbance and causes a system error. This error, which is characterized as being dynamic, was previously calculated in an engineering application problem. For this problem, it will be assumed that this error is 0.001 degrees.

Wind gusts cause servo subsystem errors in addition to antenna subsystem structural errors. This error is usually treated by separating the steady wind component from the gust portion. If the servo system is type 1 or greater, the servo action reduces the error due to the steady wind component in the steady state. The effect of the gust portion must be treated by statistical techniques. This error is calculated in an engineering application problem in Chapter 8. For this type 1 system, the resulting error due to the steady wind component is assumed to be zero, and the 3σ random error due to the random wind gusts is assumed to be 0.02 degrees.

Servo noise is caused by several sources. Notable contributors are the amplifiers, transducers such as tachometers, and the motor. This error is characterized as being random and the 3σ value is assumed to be 0.01 degrees for this system.[10]

Receiver Subsystem

Thermal noise jitter is a function of the signal-to-noise ratio in the receiver, the radar beamwidth, and the bandwidth of the closed-loop tracking sys-

Table 2.3
Error Budget for Elevation Axis of Tracking Radar

	Error Characteristic		
Source of Error	Dynamic	Random (3σ)	System-Associated
A. Antenna Subsystem			
1. Boresight			0.002°
2. Structural — Solar Heating		0.005°	
3. Structural — Gravity			0.001°
4. Structural — Wind		0.01°	
B. Mechanical Subsystem			
1. Leveling			0.01°
2. Bearing Wobble		0.002°	
C. Servo Subsystem			
1. Dynamic Lag	0.01°		
2. Nonlinear Friction	0.001°		
3. Wind Gusts		0.02°	
4. Noise		0.01°	
D. Receiver Subsystem			
1. Thermal Noise Jitter		0.01°	
2. Phase Shift		0.005°	
E. Data Unit Subsystem			
1. Encoder Readout		0.003°	
2. Noise		0.002°	
Total	0.011°	0.0275° (RMS)	0.018°

tem. In addition, the signal-to-noise ratio is a function of range. It is characterized as a random error and the 1σ value is assumed to be 0.01 degree for this system.[10]

Phase shift in the receiver also results in system error. This error is characterized as being random and the 3σ value is assumed to be 0.005 degrees for this problem.[10]

Data Unit Subsystem

The use of a digital encoder results in *encoder readout* error. This is a function of the device's resolution. Using a 17-bit encoder, a resolution of 2^{17} is achieved, which results in an angular 3σ readout error of approximately 0.003 degrees. This error is characterized as random. The system is very sensitive to *noise* in the data unit's transmission, since this device is a direct feedback element, and control systems are very sensitive to changes in the prime feedback transducer.[4] This noise can be caused by electrical pickup, any data unit gearing wobble, and distortion in the reference voltage. This error is characterized as random and its 3σ value is assumed to be 0.002 degrees for this system.

These errors are all listed in the error budget table shown in Table 2.3. The dynamic error is approximately 0.011 degrees and most of this can be corrected if a data processor is available with the tracking radar. The system associated error results in an error of approximately 0.018 degrees, which may also be corrected in the data processor. In order to combine the 3σ, independent random errors, use is made of the RMS technique discussed previously in this chapter. Applying Equation (2.38) to Table 2.3, the 3σ RMS error is found to be given by 0.0275 degrees. Combining the dynamic error, the RMS error, and the systems-associated error by means of Equation (2.42), the resultant total system error is 0.0346 degrees.

This example is very useful to illustrate the procedures to be followed in analyzing the accuracy of large, complex, systems. In addition, the reader should recognize that the system error cannot be given by merely stating one number. He should list the components of the dynamic, random, and systematic errors. In addition, he should also estimate the percentage of the dynamic and systems-associated errors which can be corrected for in a data processor, and whether it is necessary and economical to do so. Furthermore, he must specify whether the random errors are 1σ, 2σ, or 3σ, in order to completely specify the accuracy of the system.

References

1. Adelman, S. and Shinners, S.M. "Automatic Tracking Considerations for Ballistic Targets." Washington, D.C.: Fifth National Convention on Military Electronics, 1961.
2. Eckman, D.P., ed. *Systems: Research and Design*. New York: John Wiley & Sons, Inc., 1961.
3. Mishkin, I.E. and Braun, L., Jr. *Adaptive Control Systems*. New York: McGraw-Hill Book Company, 1960.
4. Shinners, Stanley M. *Modern Control System Theory and Application*. Reading, Mass.: Addison-Wesley Publishing Company, 1972.
5. Truxal, J.G. *Control System Synthesis*. New York: McGraw-Hill Book Company, 1955.
6. Shinners, S.M. *Techniques of System Engineering*. New York: McGraw-Hill Book Company, 1967.
7. National Defense Research Committee. "Parameters of Probability Distributions." *Applied Mathematics Panel Note No. 13*, June 23, 1964.

8. Machol, R.E., ed. *System Engineering Handbook*. New York: McGraw-Hill Book Company, 1965.
9. Chestnut, H. *Systems Engineering Tools*. New York: John Wiley & Sons, Inc., 1965.
10. Skolnick, Merrill I. *Introduction to Radar Systems*. New York: McGraw-Hill Book Company, Inc., 1962.
11. Mark, D.G. and Stember, L.H., Jr. "Variability Analysis," *Electro-Technology* 76 (1965).
12. Uspensky, J.V. *Introduction to Mathematical Probability*. New York: McGraw-Hill Book Company, Inc., 1937.
13. Vander Velde, W.E. "Make Statistical Studies on Analog Simulation." *Control Engineering*, (June 1960).
14. Blake, B. "Four Views on Train Control." *Control Engineering* 11 (1964) pp. 62-68.
15. Nakamura, I. and Yamazaki. "On the Centralized System for Train Operation and Traffic Control—Including Signaling and Routing Information." *Railway Technical Research Institute* 5 (1964) pp. 9-11.

3 System Reliability, Maintability, and Availability

Introduction

The increasing complexity of modern large systems has made the concept of reliability a very important factor in the overall system design. Now it is appropriate to discuss reliability, since the systems engineer must design equipment which will work in theory and practice. Although a chapter on reliability would not have been found in a book on engineering systems a decade ago, the student and practicing engineer must both be made aware of its importance in the complex technology of today.

To express reliability in quantitative terms, it is necessary to develop a mathematical model of the overall system and analyze its performance under realistic operating conditions. The generally accepted definition of reliability is as follows:[1,2]

Reliability is measured by the probability of a device performing its purpose adequately for the period of time intended under the operating conditions encountered.

Therefore, reliability can be viewed as a measure of successful systems performance when needed. To illustrate this point, consider the tracking radar problem which is concerned with accurately positioning a radar beam on a moving target. The reliability or probability of successfully tracking a target, which may last for only a few seconds or minutes, must be extremely high. Probability requirements as high as 0.99 are quite common. However, the reliability of an automobile radio receiver, used intermittently over a five-year period, could be as low as 0.5 and still give acceptable results.

A large amount of effort for improving the reliability of complex systems has been concentrated in the area of components. Use of transistors instead of vacuum tubes, integrated circuits instead of transistors, and partial redundancy techniques have all been tried. Each of these techniques will be evaluated in the following sections. In general, these methods are limited from the standpoint of the state of the art of components. It will also be shown that the concept of partial redundancy is also limited, since a chain is only as strong as its weakest link.

A very powerful tool being employed in many systems is the group redundant approach. A group redundant system combines single elements into logical functional groups which are operated in parallel. A failure detector is used at the output of each group. Its function is to switch from a normal channel to a redundant channel upon the failure of a normal channel. The group redundant approach has the advantage of a high reliability improvement factor at a relatively small increase in overall engineering effort. The application of group redundancy will be illustrated in detail.

Definitions

A few definitions, necessary for understanding the material to follow in this chapter, are presented in this section. In addition to understanding the concept of reliability, we will consider associated terminology such as probability of failure, the reliability improvement factor, the mean-time-between-failure, the mean-time-to-restore, and availability. As presented in the introduction to this chapter, the commonly accepted definition of reliability is the probability that given equipment will perform for a specified time interval under a set of specified conditions. This definition implies that the systems engineer must know the required time of operation, the operating environment, and a criterion of what is acceptable. Obviously, different operating conditions can be encountered. For example, a missile control system incorporating a tracking radar would be concerned with successfully operating over short periods of time. An air traffic control system which regulates the departure and arrival of aircraft at a large airport would be designed for reliable continuous operation over long periods of time. The extreme opposite would be a solid-propellant rocket engine of a missile. It is designed to be reliabile on a one-time basis of operation.

Experience with large electronic systems has shown that their failure characteristics follow definite patterns. A plot of the failure rate vs. time of a typical electronic component as a transistor indicates that the pattern can be segmented into three distinct types of failures:

1. Initial failures
2. Normal failures
3. Aging failures

The *initial failures* are concerned with malfunctions due to poor design; they are primarily caused by poor engineering and marginal components. They are characterized by a rapidly decreasing failure rate vs. time. The *normal failures* are associated with the random failures occurring during the primary operating life of the equipment. They are characterized by a relatively constant failure rate vs. time. The *aging failures* are caused by

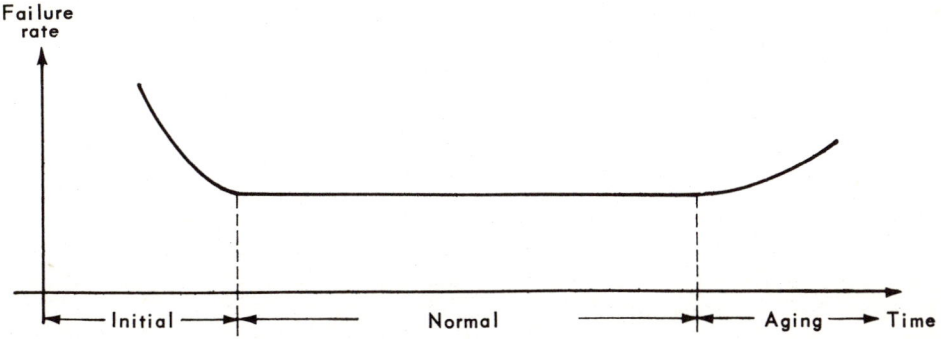

Figure 3.1. Typical Curve of Failure Rate vs. Time[2]

malfunctions due to excessive wear after expected useful design life of the equipment has been exceeded. They are characterized by a rapidly increasing rate of failure vs. time.

A typical curve of failure rate vs. time appears in Figure 3.1. Based on actual test data from large, complex electronic systems, the flat portion of the curve can be considered random corresponding to a Poisson process.[3] This characteristic is desirable from a systems engineer's viewpoint since it permits mathematical predictions of reliability based on short-term test data. The failure characteristics of a Poisson process indicate many short intervals, fewer larger ones, and extremely few long intervals. A plot of these failures forms an exponential curve.[3,10] These observations are represented mathematically by an exponential failure law. Therefore, the reliability, r, of a practical system is given by

$$r = e^{-nt/T} = e^{-\lambda t} \tag{3.1}$$

where t = cumulative operating time of system; T = mean time-to-fail of the elements; n = number of elements in the system; λ = summation of failure rates of all system elements.

In addition to reliability, the systems engineer is concerned with several other measurable characteristics of reliability. These include the probability of failure, the reliability improvement factor, the mean-time-to-restore, and the availability of a system. Furthermore, the meaning of an element, component, redundancy, and group redundancy are also useful for the discussion of this chapter. Therefore, we present the following definitions for these terms:

The *probability of failure*, p_f, of a system is defined as one minus the reliability. Its defining relationship is given by

$$p_f = 1 - r = 1 - e^{-nt/T}. \tag{3.2}$$

Assuming that $nt/T \ll 1$, a series expansion can reduce this equation to approximately

$$p_f \approx \frac{nt}{T}. \tag{3.3}$$

The *reliability improvement factor, Q*, is the ratio of the probability of failure of one system to that of another.

The *mean-time-between-failures (MTBF)* is defined as the average value of the time intervals between successive failure of equipment over the total operating time. Its relationship is given by

$$\text{MTBF} = \frac{\text{Equipment Operating Time}}{\text{Number of Observed Failures}}. \tag{3.4}$$

The *mean-time-to-restore (MTTR)* equipment is defined as the time required to locate a failure and repair it.

The *availability* of a system is defined as the probability that the equipment will be able to perform its intended function when required. Basically, it is the ratio of satisfactory operating time to the total operating time of the system. Its defining equation is given by

$$\text{Availability} = \frac{\text{MTBF}}{\text{MTBF} + \text{MTTR}}. \tag{3.5}$$

An *element* is the simplest functional part of a system. It can be mechanical, as gearing, or electronic as an amplifier stage containing one tube, or equivalent transistor, and its associated electronic *components*.

Single-element redundancy consists of two elements which operate in parallel, only one of which is required for proper operation.

Group redundancy consists of two functional groups of elements, only one of which is required for proper operation.

The Basic Foundations of a Reliability Program

What are the basic foundations of a reliability program for a large complex system? In this section, we shall concern ourselves with the tools the systems engineer has at his disposal to implement a successful reliability program. An actual program is concerned with the prevention of failures, correction of failures, and the detection and repair of the failures. To present a reliability program which is practical and has been proven, we will make use of the concepts presented by Applegate[4] which are concerned with the reliability program of a computer.

The reliability program must be a comprehensive effort from the start of the project through product delivery in order to be successful. Many important decisions must be made at the very start of the program as indicated in Chapter 1. These include trade-offs that are made between reliability and performance, schedule, cost, maintainability, life expectancy, power consumption, and weight.

Organization and Functions

The reliability group of a typical program is usually a small group of highly specialized engineers. The manager of the reliability group usually reports directly to the program manager, and must get the support of top management if he is to be effective. The functions of the reliability group manager are to specify the reliability goals, and to monitor the progress of the system from a reliability viewpoint.

In general, the reliability group has the following primary functions:

1. Collect, process and store basic reliability data on failures, successes, and operating intervals.
2. Analyze reliability data.
3. Perform reliability studies and design reviews.
4. Recommend to management corrective action for problem areas.

In addition, the reliability group should have control over the design by approving purchase drawings, drawing releases, and any changes to drawings.

Standardization

To design a reliable system, a great amount of care must go into the selection of component parts. Factors which must be analyzed are the availability of life-test data for reliability calculations, the uniformity of the manufacturing process, and the number of available sources for the part. Component selection is a highly specialized topic in which actual experience plays a very important role.

To minimize the amount of part testing and circuit analysis, a standardization program is very useful. In addition, such a program has several additional benefits. These include lower parts cost resulting from larger purchase orders, more economical testing, and simplification of the logistics problem. The reliability group should provide a standard parts manual with ratings and application data for the engineer.

Testing Component Parts and Circuits

It is extremely important to obtain dependable data on the failure rate and life expectancy of each part used in the system under realistic operating conditions. The first step in this process is to evaluate the vendor's available data. Unfortunately, most of these data are inadequate. For example, we desire failure data of each part under the expected variations to be encountered with temperature, humidity vibration, shock, pressure, and radiation. In addition, variations in line voltage and frequency should be considered. Therefore, it is usually necessary for the manufacturer to perform additional tests on components to be utilized in the system.

This type of testing is usually performed on a statistical basis to minimize its cost. In addition, commonly used circuits should also undergo these tests. All environmental tests should be conducted under the worst conditions anticipated. Analysis of this test data should be performed by computers utilizing statistical tools, as described in Chapter 7. In military work, the systems engineer should find MIC-STD 217A, titled "Reliability Stress and Failure Rate Data for Electronic Equipment" useful.

The reliability program should also have a "part review program" for the life of the project. Such an effort serves two purposes. It prevents obsolescence by taking advantage of any new developments in component parts. In addition, it permits periodic requalification of vendors' products that were originally approved.

Guidelines for the Design of Reliable Circuits

Let us assume that we are concerned with systems whose reliability is basically a function of its electronic circuitry. For example, in command and control systems, the reliability is primarily a function of its electronic equipment which includes various sensors, digital computers, electronic displays, and communication equipment. The reliability of electronic circuits is especially important. We shall consider techniques which permit high circuit reliability. Redundancy methods are discussed in later sections of this chapter.

Foremost, the circuit design engineer should *utilize reliable components* in the circuit which are consistent with performance, cost, and schedule requirements. For example, integrated circuits have markedly improved the reliability of electronic circuits. The generally accepted rule of thumb is that an integrated circuit is as reliable as one discrete component such as a transistor.[5] Therefore, by replacing n transistors and other components with one integrated circuit, one can essentially compare the reliability, r, for the replacement integrated circuit with the reliability r^n for the previous circuits, assuming that the reliability of the discrete components and integrated circuits are the same. This is a very important result to recognize. It is discussed in greater detail in the following section, on semiconductor circuits.

Another very important criterion in the design of reliable circuits is to *derate components* and use them conservatively. To accomplish this, it is important to obtain realistic failure data on component parts as a function of voltage, power dissipation, temperature, humidity, shock, vibration, etc. For example, in the case of resistors, curves are usually available which illustrate the percent failure as a function of temperature for various powers dissipated.

A third method for creating a reliable design is to allow for specified manufacturing tolerances and *design the circuit on a "worst-case" basis*. Each circuit should be evaluated under the worst-case conditions. If this analysis indicates that the circuit is not satisfactory, a statistical analysis should be performed to determine the probability of unsatisfactory performance. The basic concept of this statistical circuit analysis is to insert the component distribution data into the circuit equations to determine the probability of failure. Computer programs have been developed to perform the statistical analysis.[3]

Failure Reporting and Corrective Action

The reliability program must also incorporate a method for reporting failures and a system for accomplishing corrections. Each failure must be carefully investigated. After determining the cause, corrective action must be instituted, to prevent future occurrences. However, if the cause of the failures cannot be determined, statistical data should be recorded to indicate the areas which require further reliability tests and investigations.

Use of Semiconductor Circuits for Improving Electronic System Reliability

Examination of Equation (3.1) indicates that the reliability of a system can be improved by decreasing the ratio n/T. Basically, this means that the number of elements, n, must be minimized and the mean time to fail of the elements, T, must be maximized. As far as the number of elements in the system are concerned, good engineering practice dictates that the system should be designed with the minimum number of elements required to accomplish the functions. Therefore, from a reliability viewpoint, the systems engineer has very little control of this factor. In addition, the mean time to fail of the elements is limited by the state of the art of components. Assuming that the most reliable components are already being used, the systems design engineer has very little control over this factor as well.

The use of transistors instead of vacuum tubes will improve the reliability of a system. However, this is not the answer to the problem of increasing the reliability by two orders of magnitude. For example, even if the life of transistors were infinite, they are but one of many factors that contribute to the reliability of a device. Many studies[7] have indicated that vacuum tubes, or equivalent transistors, account for approximately 15 percent of the parts population and account for 67 percent of all failures. Therefore, other items such as capacitors, resistors, transformers, rectifiers, etc., represent 33 percent of all failures. Thus, if transistors had an infinite life, the system failure rate would still be reduced only by a factor of three. This simple analysis indicates that attempting to improve systems reliability by stressing only component reliability is severely limited by the state of the art of components.

As an example of what probability of failure the systems design engineer can expect from equivalent vacuum tubes and transistor circuits, consider a device which utilizes one hundred vacuum tubes or equivalent transistors. Assuming that the vacuum tubes are the weakest link in the chain and their mean time to fail is 7,500 hours, Equation (3.3) indicates that the probability of failure of the system is given by

$$p_{f_1} \approx \frac{n_1 t}{T} = \frac{100t}{7500} = \frac{t}{75}. \tag{3.6}$$

For the case of an equivalent transistorized system, it will be assumed that the mean time to fail of the transistors is the same as the mean time to fail of all other components, as resistors. For this case, a mean time to fail, T, of 90,000 hours is assumed. In addition, assuming that transistors account for 15 percent of all components in the device, the number of elements, n_2, which must be considered in this analysis is given by

$$n_2 = \frac{100}{0.15} = 667. \tag{3.7}$$

The corresponding value of probability of failure of the system is given by

$$p_{f_2} \approx \frac{n_2 t}{T} = \frac{667t}{90,000} = \frac{t}{135}. \tag{3.8}$$

Therefore, the reliability improvement factor, Q, which is the ratio of p_{f_1} to p_{f_2} is given by

$$Q \approx \frac{p_{f_1}}{p_{f_2}} = \frac{t/75}{t/135} = 1.8. \tag{3.9}$$

This result is very interesting, since it shows that completely transistorizing a fairly complex system improves its actual reliability by a factor less than 2. Such an engineering effort, however, is appreciable. Clearly, then, replacing vacuum tubes with transistors is not the answer to the problem of improving the reliability of a system by two orders of magnitude.

Semiconductor integrated circuits offer the potential for great improvement of systems reliability and has been a major factor in the widespread introduction of microelectronics into aerospace programs.[6] The reduction in the number of individual, discrete parts with the reduction in the associated circuit interconnections has contributed greatly to improvement in systems reliability. The reliability of a conventional circuit is predicted on the basis of the reliability of its discrete component parts. As an example,[6] consider the reliability of a NAND gate. In its discrete form, the probability of failure of the overall circuits is the product of the probability of failure of its 34 individual components which include 13 resistors, 7 capacitors, and 17 transistors. In addition, its 215 interconnections must be considered. However, a monolithic integrated circuit, performing the same circuit function, has only one probability of failure which is lower than the probability of failure of the individual transistors. Therefore, systems reliability improves in proportion to the complexity of the discrete circuit which is being replaced by the integrated circuit.

In addition, the reliability of integrated circuits is better than that of discrete component circuits due to various other factors. These include fewer circuit intraconnections, greater resistance to its environment, and circuit standardization. The integrated circuit usually has less than half the number of intraconnections of its discrete equivalent circuit. The integrated circuit also offers greater resistance to its environment due to its small weight and size. Therefore, it is less susceptible to shock and vibration. In addition, as discussed in the definitions section of this chapter, circuit standardization is an important part of system reliability. Integrated circuits improve system reliability from this viewpoint, too, since each circuit is carefully designed by trained circuit designers. Since the economics of the circuit manufacturers depends on a large-volume market for each circuit, they spend much time and effort optimizing the performance of each circuit.

Let us next reconsider the example presented earlier in this section. Originally, we found that the reliability improvement factor increased by a factor of approximately 2 when we went from a vacuum tube to an equivalent transistorized design. Let us now determine the reliability improvement factor in going from a discrete transistor component design to an integrated circuit design.

Assume that the original semiconductor discrete component circuit, which consisted of 667 overall components, can be designed utilizing 20 integrated circuits. Also assume that the mean time to fail, T, for this type of integrated circuit is 100,000 hours.[6] The corresponding value of probability of failure of the system consisting of 20 integrated circuits is given by Equation (3.3) as

$$p_{f_3} \approx \frac{n_3 t}{T} = \frac{20t}{100,000} = \frac{t}{5,000}$$

Therefore, the reliability improvement factor Q, of the integrated circuit device compared with its discrete transistorized equivalent device is the ratio of p_{f_2} to p_{f_3}:

$$Q = \frac{p_{f_2}}{p_{f_3}} = \frac{t/135}{t/5,000} = 37.04$$

This result is extremely important because it indicates that integrated circuits are very powerful tools for improving the reliability of electronic systems.

Redundancy

Redundancy techniques make use of two parallel elements to reduce the probability of failure. The principle of this concept is quite old and very simple to understand. For example, if the probability of failure of one element over a given period of time is 0.0001, the probability of failure of two redundant elements is $(0.0001)^2$, 0.00000001. Therefore, by using twice as many elements, the probability of failure has been reduced 10,000 times and the reliability improvement factor, Q, is 10,000.

The reduction in weight, volume, and cost associated with integrated circuits makes the use of full redundancy practical in some electronic systems. However, although complete redundancy appears to be a very valuable engineering tool for improving the reliability of each element, making every element redundant and providing a failure detector for every pair of elements makes this approach very unattractive from a practical viewpoint for the great majority of systems. In addition, as will be shown on pages 63-66, partial redundancy improves the overall system reliability only slightly.

How can the systems design engineer use this very powerful concept in a practical manner? The answer lies in *group redundancy*. Group redundancy consists of utilizing two functional groups of elements in conjunction with a failure detector to determine malfunctions.

The following sections consider the concept of group redundancy, as applied to practical control systems.[8] Partial redundancy is also considered and shown to be a waste of effort except for a few special cases. The very practical problem of failure detection is thoroughly considered. The systems design engineer should attempt to capture the very important concepts presented and store them in his memory for possible future systems applications.

Grouping and Failure Detection Techniques for Parallel Redundancy[8]

A practical, readily applicable approach of group redundancy is to have another duplicate channel for each channel of elements which performs a function. This results in a normal (or operating) channel and a redundant (or standby) channel. Along the problems inherent in this method are the techniques used in the failure detection and switching circuitry.

There are unique problems when applying these techniques to feedback systems: quiescently a feedback system has essentially very small null signals in the noise region, and dynamically it has transients. Other problems are introduced by the fact that a feedback system normally has both d.c. and carrier portions.

The method of failure detection and switching can utilize any one of a number of basic forms. Let us consider and analyze the following four major techniques, as applied to a feedback control system:

The first technique is to have an input to both a normal and redundant channel with a comparator type failure detector across the channel outputs. The failure detector determines which channel is faulty upon a difference across the outputs. The testing is accomplished by disconnecting the redundant channel from the normal input and feeding in a test signal. An analysis of its output response determines if it is operating correctly. If it is, the redundant channel is immediately connected into the system. If it is not operating correctly, the normal channel is left connected in the system. This type of configuration is illustrated in Figure 3.2. This technique is applicable to a wide variety of feedback systems.

The second technique utilizes failure determination by continuous monitoring of only the operating channel. For this case, if the operating channel fails, a standby channel automatically replaces it in the system. Figure 3.3 illustrates this technique.

The third technique utilizes an extraneous test signal fed through both the normal and redundant channels. This signal must not interfere with the normal operation of the system, and is of a much higher frequency than the servo control system operating frequency range. Determination of a faulty channel by the failure detector is similar to the other techniques previously mentioned. This technique is illustrated in Figure 3.4. Note the high-pass

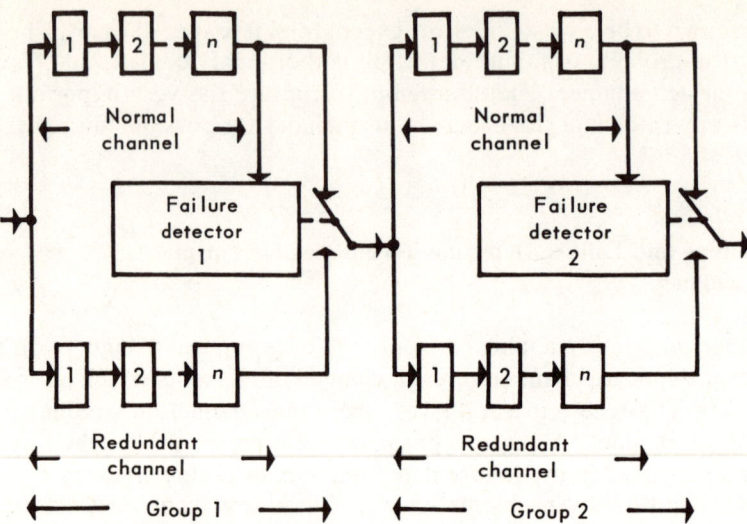

Figure 3.2. A Two-Group Redundant System Having an Input to Both a Normal and a Redundant Channel with a Failure Detector across the Channel Outputs of Each Group

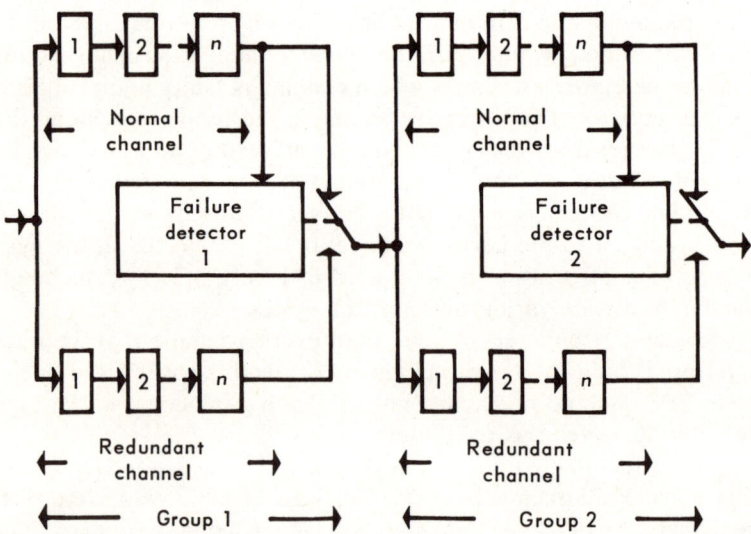

Figure 3.3. A Two-Group Redundant System Having an Input to Both a Normal and a Redundant Channel with a Failure Detector across Only the Normal Channel Output of Each Group

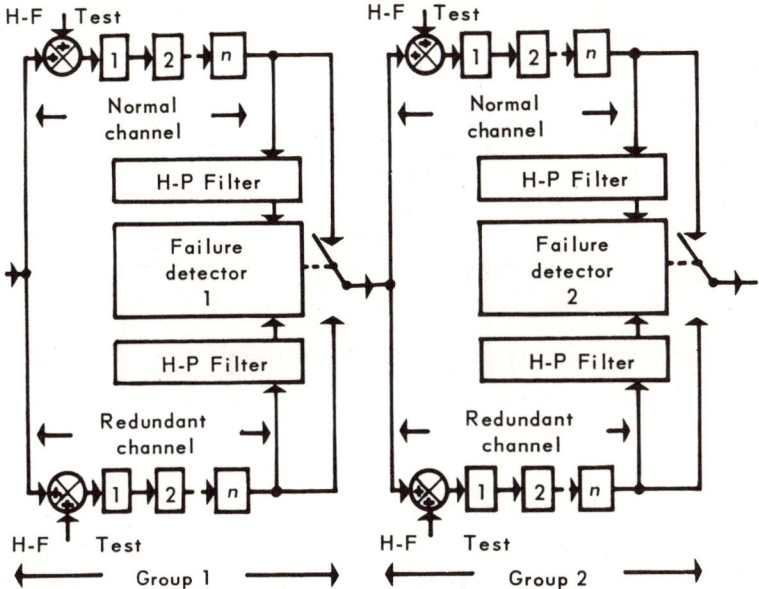

Figure 3.4. A Two-Group Redundant System with High-Frequency Test Signal Input to Both Normal and Redundant Channels

filters which allow only high-frequency information to pass to the failure detectors. This technique is applicable to a wide variety of feedback systems. However, it has certain disadvantages. The test signal acts as a disturbance signal and increases the system error. Another disadvantage of this technique is that only high-frequency failures will be determined. For example, this check would not determine if the feedback capacitor of DC operational amplifier, being used as an integrator, had short-circuited.

Other cases of interest are those where the failure detection circuitry would be more complicated than the channel under surveillance. In this case, it is advantageous to leave the determination of a failure to the decision of the operator of the equipment. An example of this is the track loop of a tracking radar which has a large amount of switching for its multiplicity of transfer functions due to its several modes of operation. Here, it is much simpler to have the operator of the equipment determine a failure, rather than perform this function automatically.

Probability Analysis of Group Redundancy

To evaluate the beneficial effects of group redundancy, it is very interesting to compare the probability of failure of the system with and without group redundancy. Consider the group redundant system shown in Figure 3.5. It

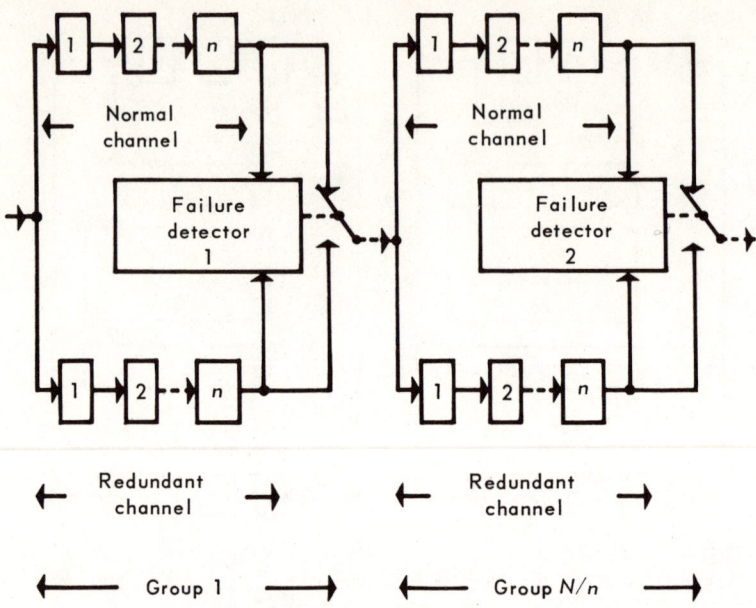

Figure 3.5. An N/n Group Redundant System Having an Input to Both a Normal and a Redundant Channel

has been shown [1,2] that the probability of failure of the ith redundant group, p_{f_i}, is given by Equation (3.10):

$$p_{f_i} = [1 - e^{-nt/T}]^2, \qquad (3.10)$$

where p_{f_i} represents the probability of failure of the ith redundant group; n is the number of elements in a redundant channel, t represents the cumulative time during which the equipment is to operate; T is the mean time to fail of an element.

This equation is based on the fact that the reliability of electronic equipment decreases with time in accordance with the exponential relationship given by Equation (3.1).

If the value of $nt/T \ll 1$, by means of a series expansion, Equation (3.11) results:

$$p_{f_i} \approx \left(\frac{nt}{T}\right)^2. \qquad (3.11)$$

Therefore, the worst-case probability of failure of a system containing a total number of capital $2N$ elements, in N/n separate redundant groups, is given by Equation (3.12):

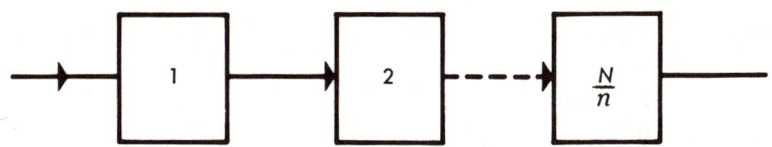

Figure 3.6. A Simple System, Containing N Elements, Without the Use of Group Redundancy

$$p_{f_s} = \sum_{i=1}^{N/n} p_{f_i}. \qquad (3.12)$$

Substituting Equation (3.11) into Equation (3.12), the probability of failure of a group redundant system, p_{f_s}, is given by

$$p_{f_s} \approx Nn(t/T)^2. \qquad (3.13)$$

To have a basis of comparison, the probability of failure which would normally be obtained without the use of group redundancy will be obtained for the system shown in Figure 3.6. The probability of failure of the ith nonredundant element is given by Equation (3.14):

$$p_{f_i} = 1 - e^{-nt/T}. \qquad (3.14)$$

The nomenclature is the same used for the previous equations. If the value of $t/T \ll 1$, by means of a series expansion, Equation (3.15) results:

$$p_{f_i} \approx \frac{nt}{T}. \qquad (3.15)$$

Therefore, the probability of failure of a system without group redundancy, containing a total of N elements, is given by Equation (3.16):

$$p_{f_{s0}} = \sum_{i=1}^{N} p_{f_i} \qquad (3.16)$$

Substituting Equation (3.15) into Equation (3.16), the probability of failure of a system without group redundancy, $p_{f_{s0}}$, is given by

$$p_{f_{s0}} \approx \frac{Nt}{T} \qquad (3.17)$$

This agrees with Equation (3.3).

A basis of comparison and evaluation is now available. By comparing Equations (3.13) and (3.17), the reliability improvement factor, Q, given by Equation (3.18) is obtained:

$$Q \approx \frac{T}{nt}. \qquad (3.18)$$

This factor will have more meaning by use of a simple practical example. The servo control system used for a naval tracking radar designed to function in a missile control system[8] had the following approximate representative constants:

$$T = 20,000 \text{ hours}$$

$$n = 15 \text{ elements}$$

$$t = 1/2 \text{ hour.}$$

Substituting these values into Equation (3.18) results in the fact that the reliability with group redundancy has been increased by a factor of 2,667. This factor indicates that group redundancy has the advantage of an extremely high reliability improvement at a relatively small increase in overall engineering effort.

Failure Detector Requirements

The failure detector requires the most critical consideration and design. There are five major requirements for a good failure deterctor of a feedback control system. These are:
 1. Fast response so that a defective unit may not damage another part of the system. For example, this is imperative in the power drive portion where a defective component could cause serious and widespread damage.
 2. Capability of detecting marginal to catastrophic failures. It is quite difficult to detect marginal failures in a feedback system since the information available to operate on is either null values, transients, or high-frequency responses.
 3. "Fail-safe" feature whereby a failure in the failure detector itself should always leave the system operating with either the normal or redundant channel. This is extremely important.
 4. Simplicity in order to increase the reliability of the failure detector. This is a guideline, in general, for the reliability of most circuits.
 5. Provisions for quickly and effectively testing the failure detector.
  Comparison in the failure detector may be accomplished by means of a simple comparator circuit as a difference amplifier, shown in Figure 3.7, or by means of pulse modulation techniques, shown in Figure 3.8. Figure 3.7 illustrates the block diagram of the failure detector circuitry corresponding to the technique shown in Figure 3.2. There is an input to both a normal and a redundant channel with a comparator type of failure detector across the channel outputs.
  Consider the operation of the circuit shown in Figure 3.7. Assume that there has been a failure in one of the channels. The difference amplifier

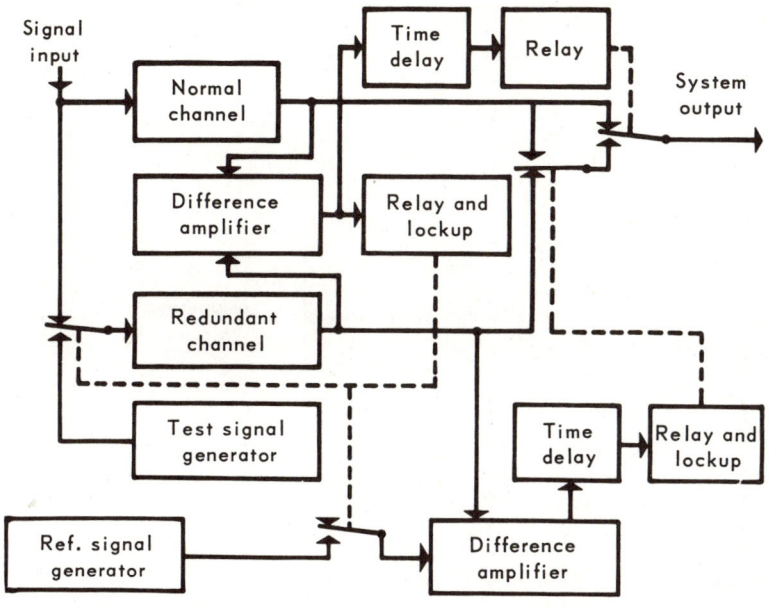

Figure 3.7. Block Diagram of a Group Redundant System Illustrating the Use of Difference Amplifiers for Detecting Failures and Checking Which Channel Has Failed.

which monitors the two channels senses the failure and trips a relay with an associated lockup circuit. This relay disconnects the redundant channel from its normal signal input and feeds it a test signal. A second difference amplifier checks the actual response of the redundant channel with that of a reference signal response. First consider the case where the redundant channel does not check out correctly. This will cause this last difference amplifier to trip a relay with an associated lockup circiut. A time delay is used in conjunction with this relay so that normal switching transients do not indicate a failure. The final portion of the failure detection system consists of the time delay and relay circuit shown at the top of the figure. This time delay circuit introduces a delay which is long enough to give the circuitry sufficient time to operate. After this predetermined time, it trips the relay shown at the system output. Since the redundant channel failed, notice that the normal channel remains connected in the system. Next consider the second case where the normal channel has failed. The difference in the functioning of this circuit is that the difference amplifier which tests the operation of the redundant channel, does not indicate a failure, and will not trip its associated relay. Therefore, the relay, which will be

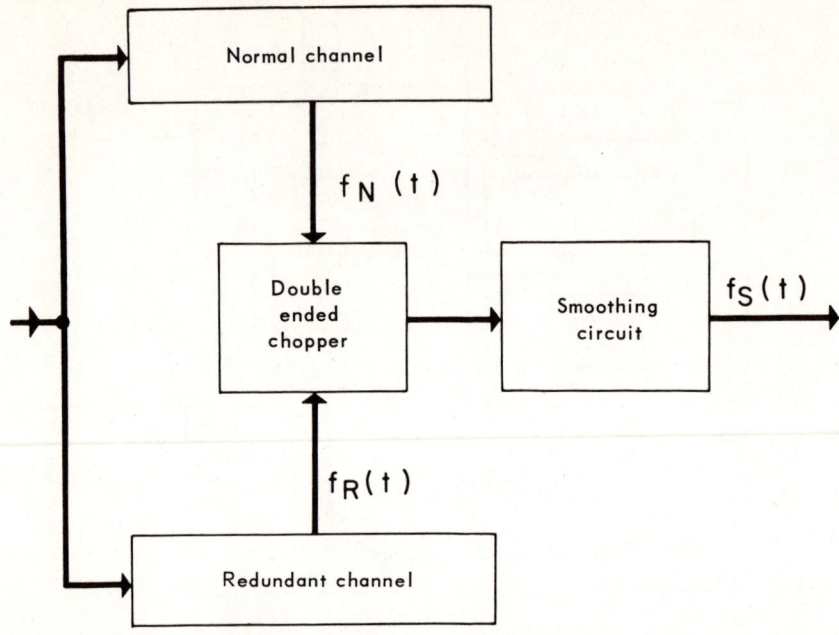

Note: When $f_N(t) = f_R(t); f_N(t) = f_R(t) = f_S(t)$. If either $f_N(t)$ or $f_R(t)$ fails, $f_S(t)$ equals the remaining signal. The smoothing circuit has a cutoff frequency above the highest component of signal frequency, and below the chopper frequency

Figure 3.8. Comparator Failure Detector Using Pulse Modulation Techniques.

tripped by the first difference amplifier discussed, and the time delay circuitry shown at the top of the figure, will disconnect the normal channel from the system, and connect the redundant channel in its place.

Figure 3.8 illustrates how one can replace the difference amplifier in the failure detection circuitry with pulse modulation techniques. In this method, the output from the modulator would have a smaller value during failures than if both channels were operating correctly. It is important that the modulation frequency be much larger than the significant high-frequency components in the transient response for adequate testing of the transient.

Figure 3.9 illustrates the group redundant approach for an entire feedback control system. A simple system, consisting of one major and one minor feedback loop, has been chosen. Failure detectors are shown at the output of the error detectors, and three other functional groups of the system. These are denoted by $G_1(s)$ in the forward portion of the major loop, $G_2(s)$ in the forward portion of the minor loop, and $H_1(s)$ in the

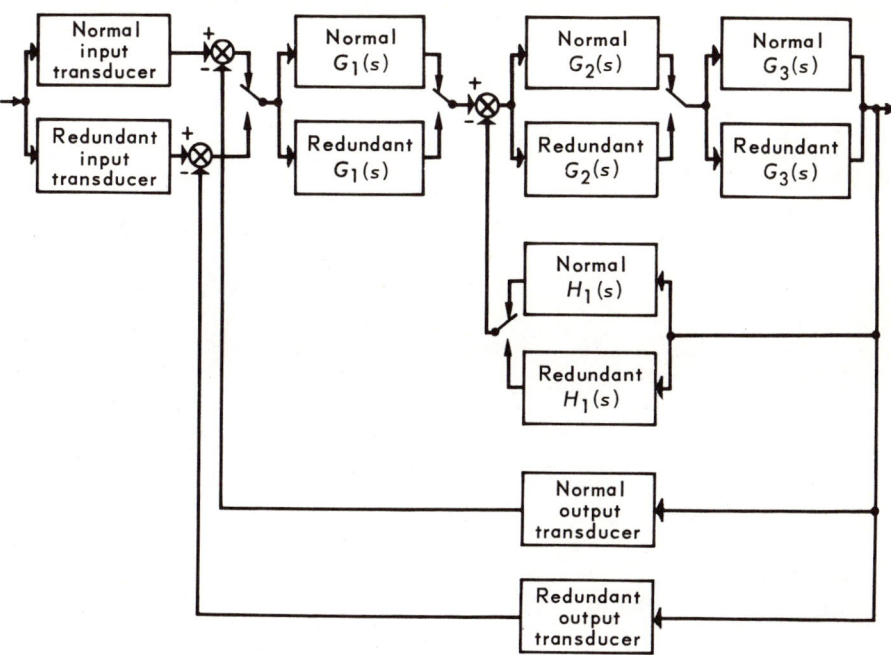

Figure 3.9. Group Redundant Approach for Entire Feedback Control System

feedback portion of the minor loop. The normal and redundant groups, labeled $G_3(s)$, have purposely been shown without a failure detector across them. In most group redundant feedback control systems, to introduce redundancy in the electromechanical and mechanical portions of the system, the servo power drive outputs are operated so that the normal and redundant channels each are operated at 1/2 rated power output during normal operation. During failures of either channel, the other channel is operated at full rated output.

Partial Redundancy

Let us now determine the effectiveness of partial redundancy. Consider the block diagram of Figure 3.10 which illustrates a partially redundant system ($n = 1$ for each element). The system illustrated utilizes N_1, nonredundant elements in series and N_2 elements in duplicate, parallel, operating redundant pairs. Obviously a chain is as strong as its weakest link, and it is obvious that the use of this technique is very limited, since a failure of any

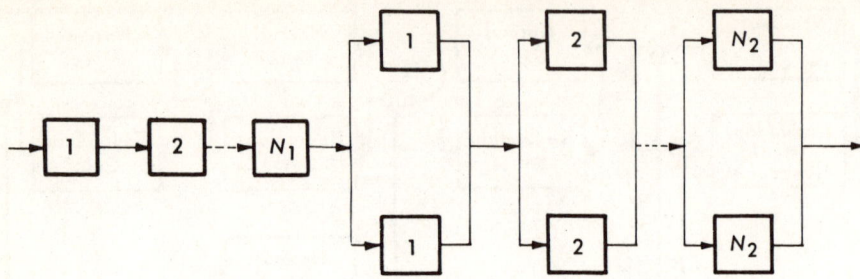

Figure 3.10. A Partially Redundant System

of the N_1, nonredundant elements immediately causes a system failure. This will be proven mathematically and the results compared with previous results.

To examine the problem mathematically, consider the probability of failure of the partially redundant system, p_{f_s}. From page 48, we know that the probability of failure of the ith nonredundant element is given by

$$p_{f_i} \approx \frac{t}{T} \tag{3.19}$$

In addition, we know from page 58 that the probability of failure of the jth redundant element is given by

$$p_{f_j} \approx \left(\frac{t}{T}\right)^2 \tag{3.20}$$

Therefore, the probability of failure of the partially redundant system, p_{f_s}, is given by

$$p_{f_s} = \sum_{i=1}^{N_1} p_{f_i} + \sum_{j=1}^{N_2} p_{f_j}. \tag{3.21}$$

Substituting Equations (3.19) and (3.20) into Equation (3.21), we obtain the following expression:

$$p_{f_s} \approx \frac{N_1 t}{T} + N_2\left(\frac{t}{T}\right)^2. \tag{3.22}$$

To determine the reliability improvement factor, Q, the probability of failure of the system without any partial redundancy, $p_{f_{s_0}}$, will also be obtained. Defining N as

$$N = N_1 + N_2, \tag{3.23}$$

the probability of failure of the system without any partial redundancy is given by (see Equation (3.17))

$$p_{f_{s0}} \approx \frac{Nt}{T}.$$ (3.24)

By comparing Equations (3.22) and (3.24), the reliability improvement factor, Q, is given by

$$Q \approx \frac{\frac{Nt}{T}}{\frac{N_1 t}{T} + N_2 \left(\frac{t}{T}\right)^2}.$$ (3.25)

This expression can be simplified to

$$Q \approx \frac{1}{\frac{N_1}{N} + \frac{N_2}{N}\left(\frac{t}{T}\right)}.$$ (3.26)

This resulting expression will have more meaning by the use of a simple practical example. Consider the naval tracking radar's control system problem discussed on page 60 where

$$T = 20{,}000 \text{ hours}$$
$$N = 15 \text{ elements}$$
$$t = 1/2 \text{ hour.}$$

Assuming that $N_1 = 7$ and $N_2 = 8$, we find that the reliability improvement factor, Q, is only 2.14. If $N_1 = 5$ and $N_2 = 10$, than Q increases to 3. For the case where $N_1 = 1$ and $N_2 = 14$, then Q increases to 15. Actually, only when the system is 99 percent redundant will Q approximate 100.

These results are interesting because they indicate that partial redundancy is extremely limited for a system having a uniform failure rate. When the system was approximately 50 percent redundant, the reliability increased by only about 2. For the case where the system was 66.7 percent redundant, the reliability improved by a factor of only 3. In the limit, where all but one of the elements was redundant, the system reliability improved only by a factor of 15. Thee results compare very poorly to the group redundant approach where the results indicate a reliability improvement factor of 2,667 for the same problem (see page 60).

The only time partial redundancy may be worthwhile is when there are a few very unreliable elements. Here, redundancy of these elements should prove beneficial to the system. However, for a system having a uniform rate of failure, partial redundancy is usually not worth the effort.

Maintainability

As discussed in Chapter 1, the maintainability of a system is a very important factor in the overall engineering of a system. Maintainability is concerned with the monitoring, checkout, corrective procedures, and preventive procedures concerned with systems operation. Since decisions affecting the maintenance philosophy directly effect the level of systems reliability that can be achieved, this discussion is included in this chapter on reliability.

The generally accepted definition of maintainability is as follows:[2]

Maintainability is the probability that, when maintenance action is initiated under stated conditions, a failed system will be restored to operable condition within a specified time.

This definition implies that maintainability is a function of the equipment's design, the operating personnel, and the support facilities. System maintainability can be improved by providing accessible test points, built-in test equipment, and built-in diagnostic aids; training the operating personnel; and providing spare parts and equipment for incorporating repairs.

Maintainability is a very important systems parameter which must be considered at the start of the program. For example, the maintenance of a system may exceed the original cost of a system. Therefore, maintainance is a major economic factor which should be recognized early in the synthesis of a system. In addition, the availability of skilled personnel is extremely limited and the performance of a system may deteriorate if the original design doesn't recognize this. Therefore, many systems incorporate automatic checkout devices and automatic failure detectors which have been discussed previously.

The time it takes to trace a failure to a particular circuit is greatly decreased by the use of built-in automatic test equipment. Modular plug-in unit construction also is a great aid. Readily accessible test points, amply provided throughout the system, also greatly decrease the diagnosis time. Clear troubleshooting diagrams also aid and improve the level of maintenance.

Adequate personnel training and motivation cannot be overemphasized. The availability of skilled personnel at operational locations is very rare. Great efforts must be made to train personnel to maintain the equipment they will operate. This will greatly improve the maintenance level and will have a great effect on system reliability.

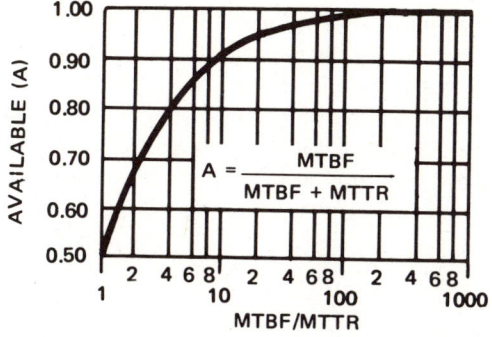

Figure 3.11. Availability as a Function of MTBF/MTTR

Reliability, Maintainability, and Availability (R/M/A)

Measures of reliability and maintainability are incorporated in availability formulations by expressing the availability in terms of MTBF and MTTR (see Equation (3.5)):

$$\text{Availability} = \frac{\text{MTBF}}{\text{MTBF} + \text{MTTR}}. \tag{3.27}$$

As discussed previously, the calculation of MTTR is related to repair hours, while the calculation of MTBF is related to component operating hours. Figure 3.11 is a graph of Equation (3.27).[9]

Figure 3.11 clearly illustrates that it is obviously desirable to keep the MTBF high and the MTTR low to keep the availability high. Observe from this figure that the availability increases as the ratio of MTBF to MTTR increases. For example, if a device has a very low MTBF, then the MTTR must be made very low to achieve a large ratio of MTBF to MTTR and a resultant high availability. On the other hand, if the MTTR is relatively high, then the MTBF must be designed to be very high in order to achieve a large ratio of MTBF to MTTR and a resultant high availability.

In the systems engineering process, a compromise should be made between MTTR and MTBF. Figure 3.11 implies that the MTBF should not be increased beyond the point where very little is gained in terms of availability. For example, an analysis of Figure 3.11 indicates that very little is gained in terms of availability for ratios of MTBF/MTTR exceeding 100.

To illustrate some practical aspects of the R/M/A interrelationship, consider an integrated circuit device which normally has an extremely high MTBF and a very low MTTR. This device is operating to the right of the

MTBF/MTTR ratio line of 1.00 in Figure 3.11. Therefore, the device has a resultant availability close to 100 percent. From a practical viewpoint, very little is to be gained in this case by designing the integrated circuit device to be repairable if special expensive equipment is required to do the job. For situations like these, therefore, it is justifiable to have a throwaway doctrine in the event of a malfunction and replace it with a new device.

References

1. Barlow, R.E. and Proschan, F. *Mathematical Theory of Reliability*. New York: John Wiley & Sons, Inc., 1965.
2. Machol, R.E., ed. *System Engineering Handbook*. New York: McGraw-Hill Book Company, 1965.
3. Sandler, G.H. *System Reliability Engineering*. Englewood Cliffs, N.J.: Prentice-Hall Inc., 1963.
4. Applegate, F.A. "Built-in Reliability." *Electro-Technology* 74 (1965).
5. Bailey, S.J. "IC Applications That Pay Off." *Control Engineering* 13 (1966).
6. *Microelectronics Device Data Handbook*. NASA Parts Publication NPC 275-1, National Aeronautics and Space Administration.
7. Chin, J.H.S. "Optimum Design for Reliability." *Sperry Engineering Review*, (March 1958).
8. Adelman, S. and Shinners, S.M. "Group Redundant Techniques," presented at the 1960 Northeast Electronics Research and Engineering Meeting, Boston, Mass.
9. *Reference Data for Radio Engineers*, 6th ed., Chapter 43. New York: Howard W. Sams & Co., Inc., 1975.
10. Ireson, W. Grant, ed. *Reliability Handbook*. New York: McGraw-Hill Book Company, Inc., 1966.

4 Management Control of System Schedule and Cost

Introduction

"Time is money." This simple, well-known statement vividly illustrates the relationship between schedule and cost. As discussed in Chapter 1, schedule and cost are two very important considerations in the overall engineering of a system. Since they are very closely related, they are considered together in this chapter.

What can we say regarding schedule and cost? Obviously it is desired to engineer the system in the shortest amount of time and with the lowest cost. This is the goal but it is not the entire story. From the systems engineer's viewpoint, schedule and cost considerations involve the use of tools to control these factors, their relationship upon the rest of the system, and what can be done in a feedback manner to adjust other system parameters if schedule and cost deviate from the desired goals.

It is very rare to include schedule and cost factors in a systems engineering volume. To the best of my knowledge, this is the first time that a full chapter is devoted to economic considerations in a systems engineering book of this type.

Why do these factors merit this much attention? Today's engineers, involved in the design of very large, complex systems in a competitive and free society must understand the relationship of performance, reliability, schedule, cost, maintainability, life expectancy, power consumption, and weight in order for their organizations to be competitive and successful. By our standards of measurement, it is necessary to produce a product that works reliably, within the allotted schedule, at a profit.

In this chapter schedule and cost are presented from a different viewpoint than economics books usually present those subjects. Since this volume is designed primarily for engineers and not accountants, we are most interested in developing the subject using tools available to the engineer. For example, we shall rely on flow diagrams, control theory, computers, simulation, and modeling as our tools for controlling cost and schedule as opposed to balance sheets. Such modern schedule control tools as PERT (Program Evaluation Review Technique) are discussed in this chapter.

This chapter also considers the operation of an engineering organization as a dynamic control system. It presents modern management theory which treats an organization's structure as a feedback control system where all the tools of control systems analysis and synthesis are applicable. The discussion demonstrates the application of feedback control theory to engineering organizations in a manner similar to its application in such varied control problems as the industrial process control and weapons control fields.

The modern engineering organization is treated as a complex machine in which block diagrams and transfer functions provide a mathematical model for performance analysis. A very important performance relationship is output vs. time. Accomplishments measured in terms of man-months of effort, cost to produce, and total time to complete the program are important yardsticks of performance analysis. Methods for reducing an organization's response time are indicated. This permits increased return on assets by making greater use of facilities over a given period. Return on assets is a very important practical criterion used for the performance of many modern organizations.

The dynamic analysis of engineering organizations was conceived of in References 1 to 3 on page 97. It represents a very practical powerful tool for program control in a very realistic technological sense. Actual case histories are illustrated throughout the chapter to supplement the presentation.

The Economic Flow Graph

A fundamental understanding of free business enterprise is very important for the evolving discussion of schedule and cost control. The economic flow graph provides a vehicle for illustrating the basic principles involved. It traces the flow of cash from the inception of a business enterprise to the sale of the finished product. By means of this flow graph, the systems engineer can vividly understand his importance in the overall daily economic life of a business.

The economic flow graph of a hypothetical business enterprise is illustrated in Figure 4.1. The issuance of stocks, bonds, and notes represents the input to the business. These contracts, or agreements, are converted into cash or credit. The next step consists of taking some of these assets and converting them into equipment, materials, and services by procurement in order to produce a product. Some of the cash or credit may be set aside as a reserve to be used for the production of different products, acquisitions, contingencies, etc. The following phase consists of converting the procured equipment, materials, and services into a product by manufacturing. During this phase, we find that the value of the finished product is composed of materials specifically purchased to produce this product, equipment usage charged off to the product in the form of depreciation, and the

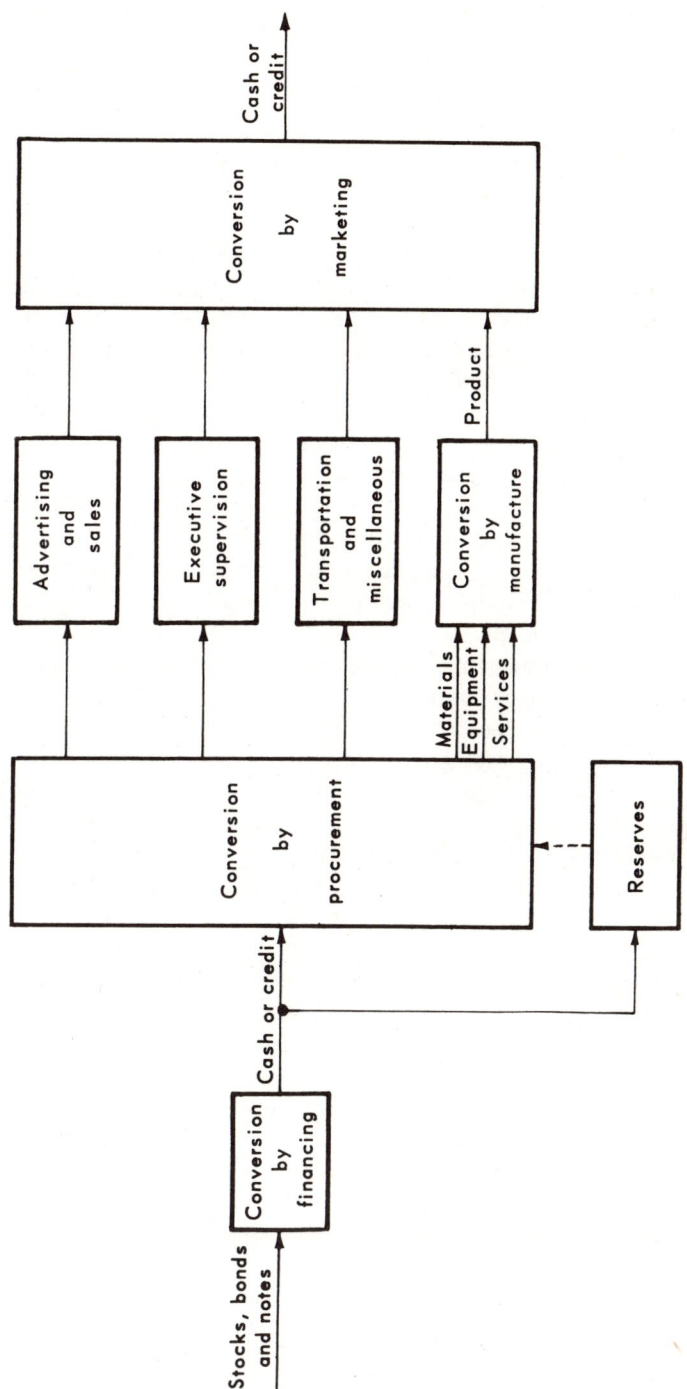

Figure 4.1. The Economic Flow Graph

services purchased to produce the product such as wages, salaries, power, insurance, and taxes. The final step in the economic flow graph consists of marketing the product for cash, credit, or notes. Associated with the cost of the product are the equipment and services charged for executive supervision, advertising, salesmen, transportation, and for other miscellaneous expenses. These cost factors are shown as parallel paths to the manufacturing path in the economic flow graph.

The overall problem has been simplified by not providing a separate path for engineering research and development. Assume that the associated equipment, material, and service costs for engineering research and development are combined with the production costs to manufacture the product. This is done in some corporations, but not in all cases. Actually if the manufacturing and engineering overheads are significantly different, it is wise to separate engineering and production costs, and to show separate paths for engineering and production.

All free business enterprises have the same basic economic problem of converting from one form of value into another form of value. Although the relative structure of the economic flow graph varies among businesses, the primary and basic problems lie in the four stages of value conversion: financing, procurement, manufacture, and marketing. How does the systems engineer who is managing a program fit into this picture? His problem usually focuses on all stages of the business's economic flow, except that of financing. However, in some very small businesses, this too is his problem. Therefore it is very important that the systems engineer understand the economic flow of values and their conversions from one form to another.

By virtue of the systems engineer's knowledge of block diagrams and feedback control theory, he can make an important contribution to the control of cost. As discussed in Chapter 1, cost control can be considered from a feedback viewpoint. Figure 4.2 illustrates such a feedback system. Desired cost is compared to anticipated cost. Logic elements are utilized for varying the specified control system requirements and subsystem characteristics, and determining their overall effect on cost. As discussed in Chapter 1, the elements labeled "specify control system requirements" and "specify characteristics of subsystem" are also common to the performance feedback path. In addition, the reliability feedback path affects the response of the elements labeled "change in cost due to control system requirement variations" and "change in cost due to control subsystem characteristics variation".

Program Management

The function of program management in the overall scheme is to unify the specialties of engineers, technicians, administrators, production personnel, and accountants to produce a product or service on schedule and at a

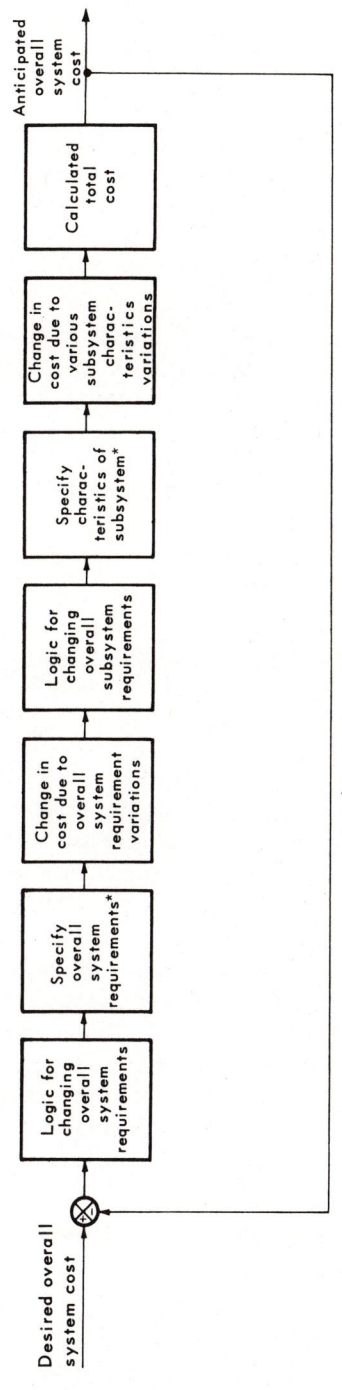

Figure 4.2. Cost Feedback Path

profit. It places together under a single command all of the technologies, skills, and resources required to make the program a reality. Fundamentally the basic problem confronting the systems manager is that of running a business within a business.

The program is usually organized vertically by tasks to be performed in the overall system, in contrast with the horizontal organization of the traditional alignment.[4] At the first level of program management is the systems manager. His function is analogous to that of a chief executive in a business. The systems manager must plan the program to satisfy the needs of the customer and to direct the program so that a product or service is successfully delivered on schedule at a profit. Systems management is concerned with the allocation of resources and the task of engineering the system. The manager must determine the proper relationships among the value of the contract, the organization's capital, the desired profit, and the technological capability of the organization. The systems engineering manager must translate the complicated and interrelated requirements of the master plan into the language of the systems designers. The actual design of the product or the performance of a service, and administrative control is accomplished at the third level. Figure 4.3 illustrates the organization of a typical system division.

As discussed in Chapter 1, every program has a multitude of time-sequential phases which involve information feedback flow. Basically, these phases consist of defining the problem, providing analytical solutions, mechanizing the equipment, and verification. The program definition determines the customer's needs. The second phase describes the system which fulfills these needs and determines the resources which must be allocated to achieve it. The mechanization phase is concerned with the assembly of the system. The final phase, verification, is concerned with testing the system to verify that it will perform its intended function in the environment in which the customer intends to utilize it.

Although an individual may provide the systems engineering management in relatively small programs, hundreds of people and many groups may be involved in very large programs. In large, complex systems, an individual or a group cannot perform these varying systems engineering management functions throughout the life of the program. For these cases, it is important that the program manager anticipate transitions between the various program phases and adjust his lower levels of management accordingly.

Time Models—PERT

The use of time models to analyze a program's schedule is of great use to the systems engineer. The concept of PERT (Program Evaluation Review Technique) has proven to be a very valuable tool for schedule control.[5-8]

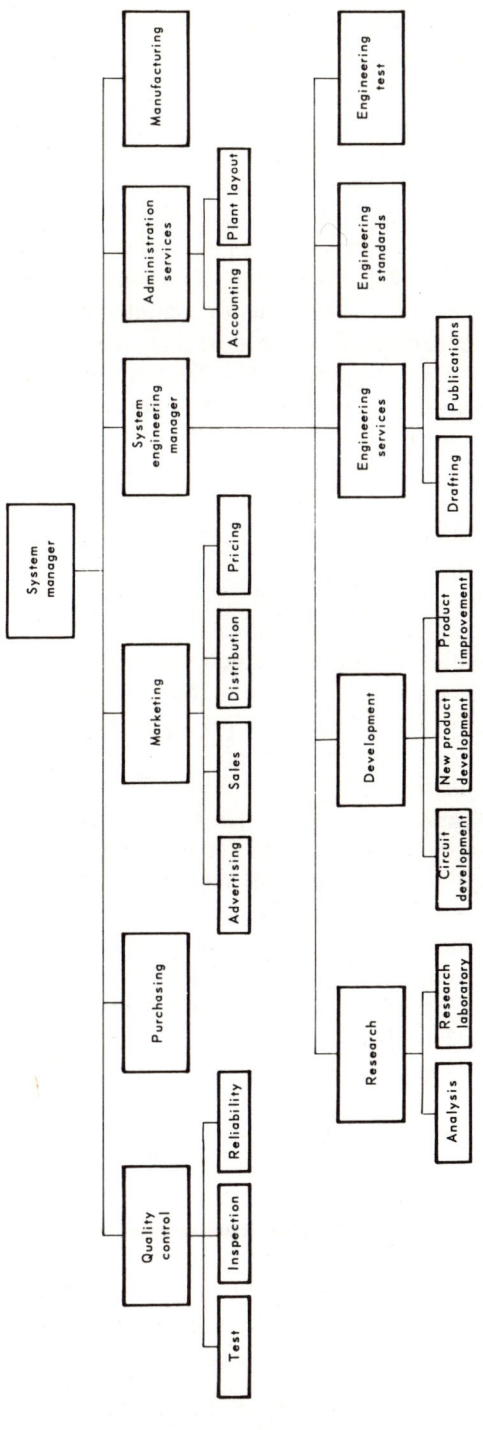

Figure 4.3. Organization Chart of a Typical Systems Engineering Division of a Large Company

By applying this technique, the program manager can establish a program plan, monitor the progress of the effort, review the status, determine items that are critical from a schedule viewpoint, and make decisions concerning adjustments. With the aid of PERT, delivery schedules of modern complex systems have been accurately predicted and controlled.

PERT was developed by the U.S. Navy in 1958 in an attempt to speed development of their Fleet Ballistic Missile System (Polaris). It was a very important tool in helping the Navy save approximately two years in the overall development time.[7] This technique has evolved to the point where it is now a requirement in many government programs. PERT is also being utilized widely in private industrial programs. It has attained this important status because of its many benefits which include integrated planning; improved communications; integration of schedule, cost, and performance data; and its viability for utilizing alternative solutions.

Nomenclature and Definitions

The PERT concept is very simple and, in many respects, is similar to signal flow graphs used by control systems engineers.[9,10] PERT is basically a *flow diagram* procedure which illustrates symbolically a plan of action required to achieve the final objective of the program. It depicts all significant events and activities necessary to complete the program. Each path in the flow diagram represents work that must be accomplished. Therefore, before generating a PERT diagram, all work activities must be clearly defined.

PERT consists of three basic phases. The first is to develop the network of events. An estimation of the time to accomplish each activity is performed in the second phase. The final phase consists of analyzing the schedule problems involved in the program.

The terms *event* and *activity* have been utilized in designing the PERT time model. An event is defined as a specific accomplishment occurring at a specific time which aids in meeting the final objectives. An activity is considered to consist of an effort which links two successive events. Figure 4.4, which illustrates a general PERT network, indicates the meaning of event and activity.

PERT subdivides a program into a number of functional subtasks which can be clearly identified. The estimated *shortest time*, *most likely time*, and *longest time* to accomplish a task are shown on the path line entering the various blocks. In addition, the *statistically expected time* to complete this task is shown below these three values along the path line. Analysis of such a time model permits identification of the *most critical path* which represents the longest time to complete the program. This path is usually shown dotted.

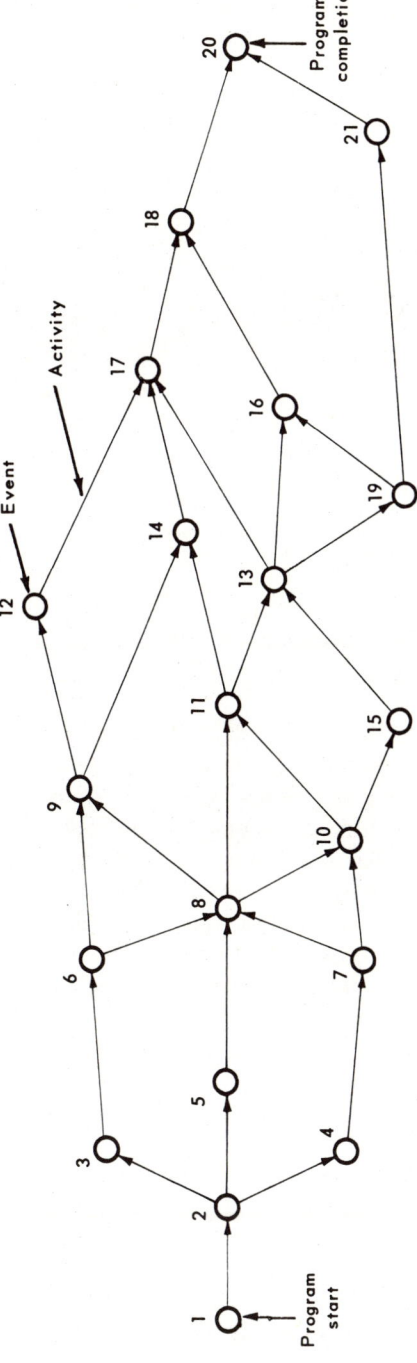

Figure 4.4. A General PERT Network

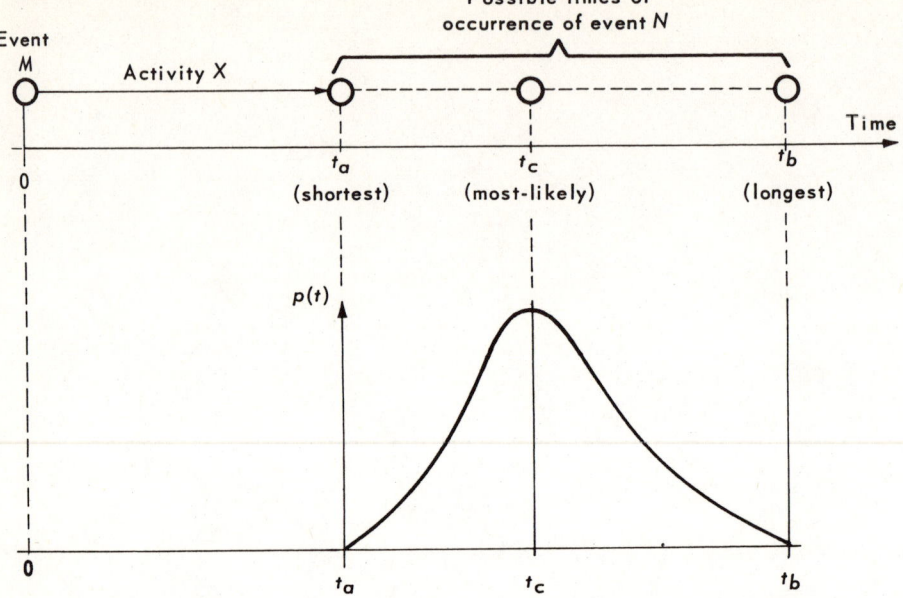

Figure 4.5. Probability Distribution Function: General Characteristics

The shortest time estimate, t_a, represents the most optimistic estimate. It assumes that everything goes well, everyone does his job correctly the first time, there are few personnel absences, and communications flow smoothly. The longest time estimate, t_b, represents the most pessimistic estimate. It assumes that everything proceeds poorly, many errors are made, there are many personnel absences, and there is a breakdown in communications. The most likely estimate, t_c, assumes the organization functions in a normal manner, there are the usual number of problems, an average number of personnel are absent, and communications between different groups are normal.

Statistical Characteristics of Time Estimates

The characteristics of these time estimates are analyzed by means of a simple example. Consider the estimate of completing activity X, which progresses the program from event M to event N. The shortest time, longest time, and most-likely times are denoted in Figure 4.5 by the notations t_a, t_b and t_c, respectively. In addition, the general characteristics of the corresponding probability density distribution, $p(t)$, are illustrated.

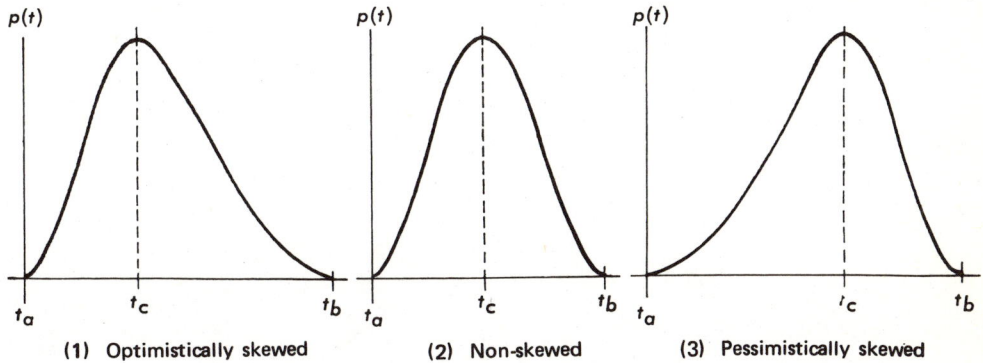

(1) Optimistically skewed (2) Non-skewed (3) Pessimistically skewed

Figure 4.6. Probability Distribution Function: Characteristic Variations

Figure 4.6 illustrates three possible probability density distributions of time estimates. The first curve is skewed toward the most optimistic time t_a. A nonskewed distribution is illustrated in the second curve. The third curve shows the case where the probability distribution is skewed towards the most pessimistic time t_b.

Derivation of the Statistically Expected Time

The mean or *statistically expected time*, t_e, represents a standard deviation calculated by one-sixth the difference of the shortest and longest time range estimates. PERT assumes that the probability density distribution of the statistically expected time can be represented best by the beta distribution illustrated in Figure 4.7. This model assumes a non-Gaussian probability density distribution of activity times. Furthermore, we assume that the standard deviation, σ_t, is one-sixth of the range:

$$\sigma_t = \frac{t_b - t_a}{6}. \quad (4.1)$$

The probability density of the beta function is given by[11]

$$p(t) = K(t - t_a)^\alpha (t_b - t)^\gamma \quad (4.2)$$

where K = constant; t_a = shortest time estimate; t_b = longest time estimate; and α and γ are functions of t_a and t_b.

A specific beta distribution can be determined if the exponents α and γ are known. The shortest and longest times, t_a and t_b, represent the two end points of this beta distribution as illustrated in Figure 4.7.

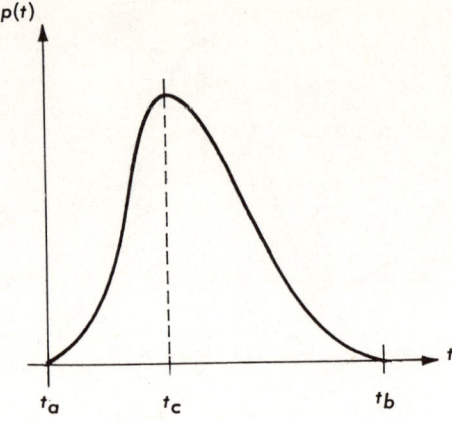

Figure 4.7. The Beta Probability Density Distribution

To reduce this probability density distribution function, the random variable x is introduced and is defined as follows:

$$x \equiv \frac{t - t_a}{t_b - t_a}. \qquad (4.3)$$

The probability density distribution of x is given by[11]

$$f^*(x) = [B(\alpha + 1, \gamma + 1)]^{-1} x^\alpha (1 - x)^\gamma, \ 0 \leq x \leq 1 \qquad (4.4)$$

Let us define the value r by the following expression:

$$r \equiv \frac{t_c - t_a}{t_b - t_a}. \qquad (4.5)$$

Denoting $E(x)$ as the mean or statistically expected value of x, and $V(x)$ as the variance of x, the following expressions can be obtained from these equations:

$$r = \frac{\alpha}{\alpha + \gamma}, \quad (\alpha = t_c - t_a, \gamma = t_b - t_c) \qquad (4.6)$$

$$E(x) = \frac{\alpha + 1}{\alpha + \gamma + 2}, \qquad (4.7)$$

$$V(x) = \frac{(\alpha + 1)(\gamma + 1)}{(\alpha + \gamma + 2)^2 (\alpha + \gamma + 3)}. \qquad (4.8)$$

The variance of t can be obtained by squaring the standard deviation which was defined by Equation (4.1):

$$\sigma_t^2 = \frac{(t_b - t_a)^2}{36}. \tag{4.9}$$

Therefore, the variance of x is 1/36. By substituting 1/36 for $V(x)$ in Equation (4.8) and eliminating γ in Equations (4.6) and (4.8), the following expression is obtained.

$$\alpha^3 + (36r^3 - 36r^2 + 7r)\alpha^2 - 20r^2\alpha - 24r^3 = 0. \tag{4.10}$$

The value of the statistically expected time, t_e, can be calculated from the transformation equation between t and x:

$$t_e = t_a + (t_b - t_a)E(x). \tag{4.11}$$

An exact plot of $E(x)$ vs. r, which can be obtained from Equations (4.6), (4.7), and (4.8) is very close to a linear relationship that can be represented by the following simple equation:[11]

$$E(x) \approx \frac{4r + 1}{6}. \tag{4.12}$$

This linear equation is accurate enough for most engineering applications, and eliminates the necessity of solving the cubic relationship given by Equation (4.10).

From Equations (4.5), (4.11) and (4.12), the relationship of the mean or statistically expected time can be obtained. Its value is given by

$$t_e \approx \frac{t_a + t_b + 4t_c}{6}. \tag{4.13}$$

The approximate relationship of Equation (4.13) is commonly used in practice for PERT applications. This alleviates the need for the solution of a cubic Equation (4.10).

The mean and variance values are sufficient to define a Gaussian curve (see Chapter 2), but are not sufficient in the case of a beta distribution. However, in PERT applications, it is convenient to denote the activity time in terms of the Gaussian distribution density also. Therefore, we obtain the variance as being the square of the assumed standard deviation (see Equation (4.9)) and the value of the mean to be represented in terms of the linear approximation given by Equation (4.13).

In order to understand the physical meaning of Equation (4.13), let us rearrange this equation:

$$t_e = (1/3)\left(2t_c + \frac{t_a + t_b}{2}\right). \tag{4.14}$$

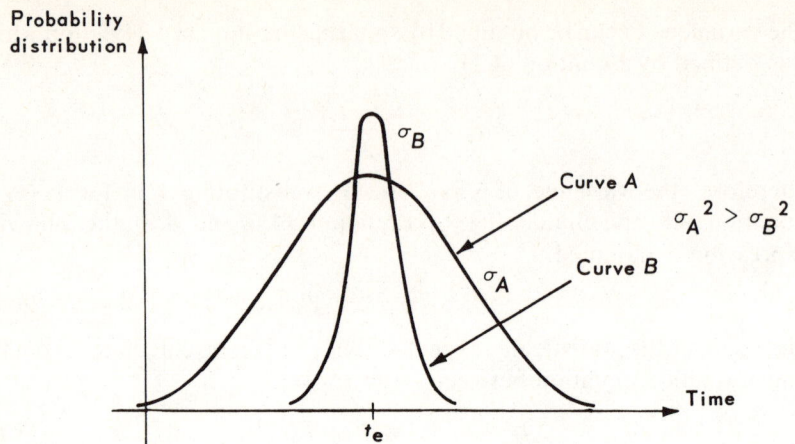

Figure 4.8. Comparison of Probability Distributions for Two Gaussian Time Functions Having Different Variances

Equation (4.14) indicates that t_e is the weighted mean of t_c and the mid range of $(1/2)(t_a + t_b)$, where the weighted coefficients are 2 and 1, respectively.

The variance of t_e is a measure of confidence in this estimate. If the variance is small, then the confidence at t_e is high. Similarly, if the variance is large, then the confidence at t_e is low. As an example, let us consider the two situations illustrated in Figure 4.8 where the probability distributions of the time estimated result in normal Gaussian distributions. The variance is related to the spread of the curves A and B shown. Since the variance of curve A is greater than that for curve B, the spread of curve A is larger than that of curve B. Therefore, the confidence in t_e from curve B is greater than that for curve A.

The *statistically expected time to complete the program*, T_E, is obtained by adding the various values of t_e from the starting event, through all paths, to the program's completion. Where two or more paths merge into a single event, the highest value of T_E is chosen.

An Example

Figure 4.9 shows a PERT model for producing engineering drawings to manufacture a special purpose computer. The time values illustrated are in weeks. This PERT diagram is concerned primarily with the engineering effort and is part of a larger overall PERT diagram which includes the preparation of manufacturing drawings, manufacturing, quality control, system test, reliability and logistics. In addition, this special purpose analog computer is part of a larger weapons control system for which a PERT time model can be implemented. The blocks in Figure 4.9 indicate major milestones for completing this task. These include the preparation of the block diagram, design, breadboard test phase, preparation of the schematic, and cabinet layout drawings. The critical path, shown dotted, encompasses the path from preliminary study, preliminary block diagram, interface specification preparation, preliminary design, preliminary breadboard test, final breadboard test, final schematic preparation, to the completion of the engineering drawings. The statistically expected time to complete the program, T_E, is 34 weeks. The shortest estimated time to completion is 23 weeks, the most likely estimated time is 33 weeks, and the longest estimated time is 41 weeks. Program schedule is usually based on the most likely estimated time, and the statistically expected time is used as a check.

An analysis of the critical path is very important for determining whether the desired schedule can be met. Every effort must be made to investigate the details of each block along this path, even going to the extent of drawing a PERT diagram for each of these subefforts. If it is impossible to meet the desired schedule, it may be necessary to follow a different plan, such as subcontracting part or all of this subsystem.

Slack Time

It is possible that a reevaluation of the efforts along the critical path will result in a shortening of its time. This is possible by bringing into play more resources along the critical path, by having more activities going on simultaneously and/or by compromising some design goals. The reassignment of resources from one activity to another activity in the critical path is determined by analyzing the PERT graph for *slack time*.

As an example of slack time, consider the simple PERT network shown in Figure 4.10. In this network, there are 4 events and 4 activities occurring. The statistically expected times to complete each activity are indicated in months. The critical path for this network is path 61-62-64 where 9 months are required to go from event 61 to event 64. By relocating men and equipment assigned to the activities along path 61-63-64 and using them for critical path 61-62-64, it is theoretically possible to save two months. The saving of two months in this simple example indicates the potential value of PERT in a network which may contain thousands of activities.

Consider the determination of slack time for the PERT network considered previously in Figure 4.9. To go from the effort labeled "preliminary design" to that labeled "final breadboard test," there are two paths. One

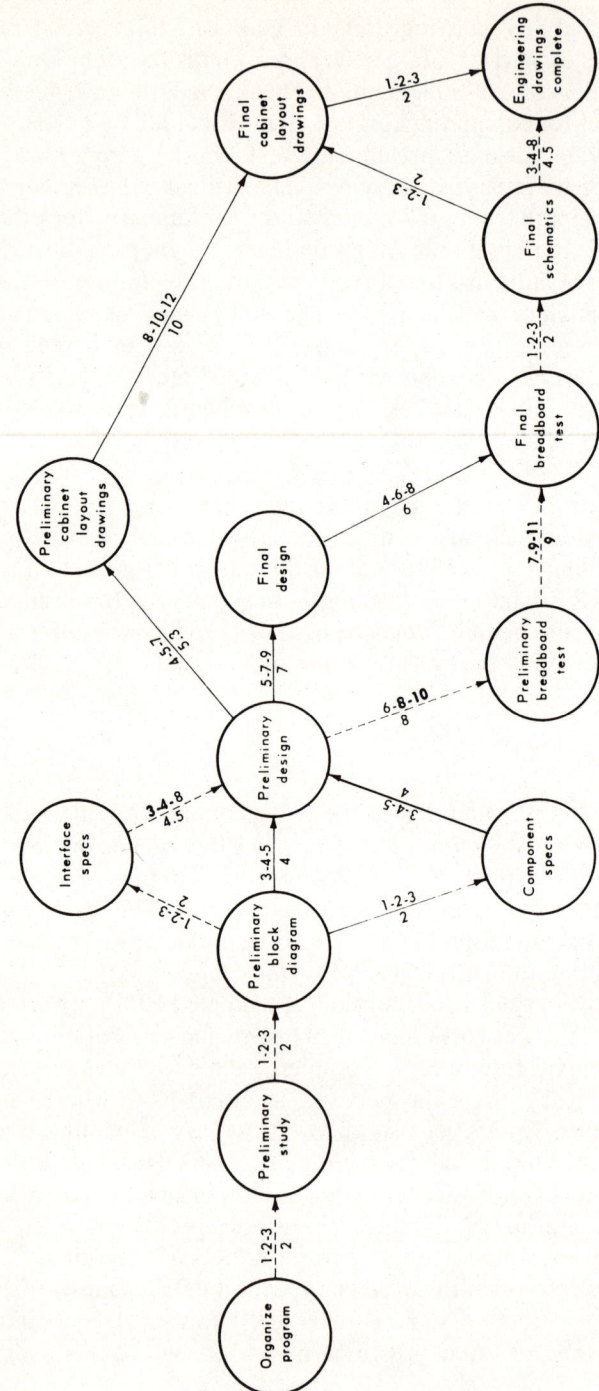

Figure 4.9. PERT Diagram for Preparing Engineering Drawings to Manufacture a Special-Purpose Computer

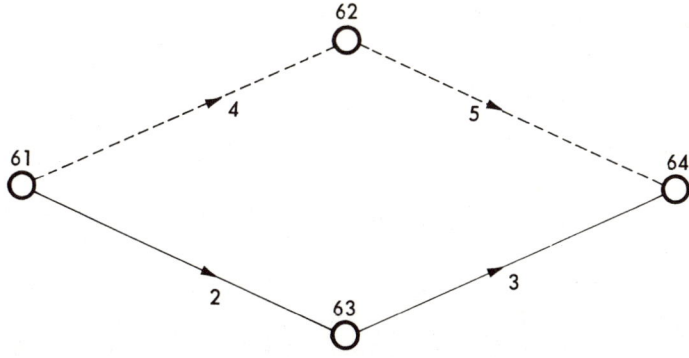

Figure 4.10. Example for Determining "Slack" Time

path has a statistically expected time of 13 weeks and the other, which is part of the critical path, has a value of 17 weeks. This results in the first path having a "slack" of 4 weeks. By relocating engineers assigned and equipment from the first path to the second path, we can foresee a saving of perhaps two weeks by equalizing both paths to 15 weeks. This has the effect of reducing the critical path time by two weeks.

Accuracy of PERT[12]

Two types of errors are introduced due to the assumptions made in the mathematical development of PERT. The first error category is concerned with the individual activities which comprise a PERT network. They are caused by the assumption that the three time estimates t_a, t_b, and t_c are fitted into a beta distribution whose standard deviation is one-sixth of its range. The second type of error is caused by assuming that the mean and variance of an overall program's distribution are the sums of the means and variances of all the individual activities which occur along the critical path of a PERT network.

An analysis of the individual activity errors indicates that basic errors are introduced by the assumption that each activity behaves as a beta distribution. Even if we assume that all activity estimates have this characteristic, then there are errors associated with the estimates of t_a, t_b, and t_c. If we go one step further and assume that t_a, t_b, and t_c are accurate estimates which do fit into a beta distribution, then the assumptions that the standard deviation is given by

$$\sigma_t = \frac{t_b - t_a}{6}, \qquad (4.15)$$

and the most likely estimated time is adequately represented by the linear approximation given by

$$t_e \approx \frac{t_a + t_b + 4t_c}{6} \tag{4.16}$$

introduce errors.

When we assume that the mean and variance of an overall program's distribution are the sum of the mean and variance of all the activities which occur along the critical path, then errors introduced are dependent on several factors. The relative number of series and parallel paths in a network, the relative lengths of various series and parallel paths, and the frequency of occurrence of common path junctions appearing along parallel paths all affect the accuracy of the model.

To improve the accuracy of PERT, the program manager should initially investigate alternative schemes for meeting the desired schedule. This can be done using simple PERT diagrams which contain relatively few activities and events. After a basic strategy has been agreed upon, further detail can be added to the PERT diagram to improve its accuracy.

Mathematical Models of Engineering Organizations

Modern engineering organizations are complex systems composed of several divisions, departments, sections, and groups. Efficient program control is obtained by means of functional capability and proper feedback control. As shown in Chapter 1, an organization's operation can be viewed as a multiple feedback loop control system composed of performance, reliability, schedule and cost loops. By virtue of feedback, a continuous comparison of the actual and desired system parameters can be made and corrective action can be taken to reduce the difference between the two.

The major elements of an engineering organization consist of engineering, drafting, and manufacturing. In addition, to complete the organization, various service groups are required. These usually include marketing, sales, quality control, reliability, and logistics. Each of these specialized subsystems has its own departments, sections, and groups geared to their specialized activities. Each of these units has certain dynamic response characteristics which can be represented mathematically by means of transfer functions.

The performance criteria used to measure accomplishment consists of measuring the relationship between output and time. The output can be in the form of man-months of effort, cost to produce, and/or total time to complete the program. As an example, let us consider an engineering program which requires 1,000 man-months to complete in a 15-month

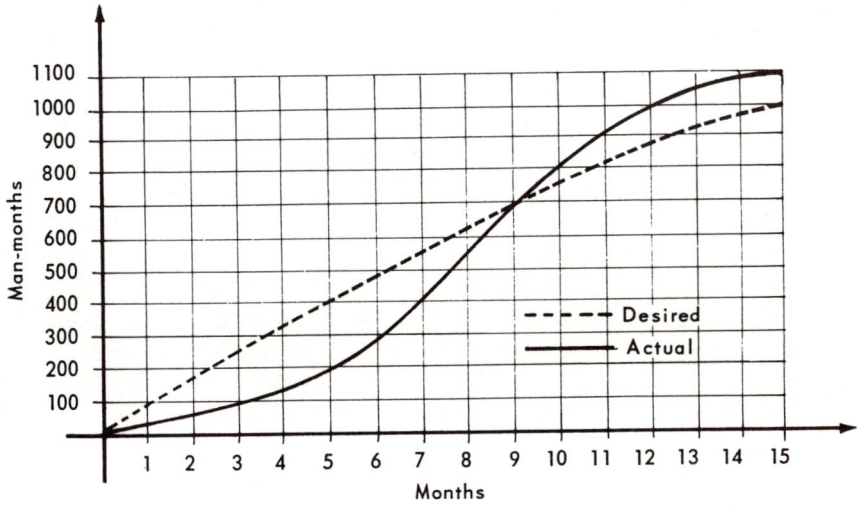

Figure 4.11. Organizational Performance

interval. The dashed curve of Figure 4.11 indicates the planned cumulative man-months expenditure. The solid curve indicates the actual cumulative man-months expenditures. Why do they differ? The differences are caused by undesirable disturbances which cause the system to deviate from the desired response. The most common causes for these disturbances are human errors, indecision, lack of material, and/or poor communication. These disturbances can affect the performance, cost, and schedule of the product. The solid curve of Figure 4.11 illustrates how these disturbances increased the manpower expenditure by 10 percent.

To investigate this type of response in further detail, the dynamic characteristics of the engineering organization's elements are represented mathematically. Studies of typical response curves of an engineering organization has indicated that they can be represented by corresponding transfer functions which consist of combinations of quadratic factors, pure integrations, pure differentiations, lag time constants, lead time constants, and transportation lags.[2,3] The characteristic of a transportation lag is significant, since it represents a delay, or dead time. It is quite common and can be caused by several factors. For example, if the engineering phase is incomplete, the production operation cannot proceed. Other common causes of time delay are indecision, lack of parts, lack of manpower, delays in transportation, etc.

The predominant mathematical term of the overall performance of a organization is usually a transfer function which is a simple quadratic lag term whose transfer function model is given by[2,3]

$$\frac{C(s)}{R(s)} = \frac{\omega_n^2}{s^2 + 2\zeta\omega_n s + \omega_n^2}, \qquad (4.17)$$

where ω_n = natural frequency of the program; ζ = variable damping factor imposed on the program by management; and $\tau = 1/\omega_n$ = time constant of the program.

This transfer function, which is the overall transfer function commonly encountered of a type 1 second-order control system, can be used to represent the mathematical behavior of the desired program's performance that was illustrated in Figure 4.11.[13] To meet the program's schedule and cost goals, the program management attempts to obtain a critically damped system where $\zeta = 1$. Deviations from this value result in an underdamped engineering operation, $\zeta < 1$, or an overdamped engineering operation, $\zeta > 1$. An underdamped operation usually results in overexpenditure of manmonths and an overdamped operation usually results in a slippage of schedule, as illustrated in Figure 4.12. The underdamped and overdamped cases are the result when program supervision is not properly and timely applied.

The overall performance of an engineering organization is dependent on the performance of several suborganizations. Let us focus our attention on a hypothetical engineering department composed of five sections:

1. Preliminary engineering analysis
2. System synthesis and design
3. Subsystem development
4. Experimentation
5. Final engineering design

Assume that the transfer function (performance) of each section is of the form given by Equation (4.17). Figure 4.13 illustrates the mathematical model used to denote the performance of this engineering department. The overall transfer function of this model is given by

$$\frac{C(s)}{R(s)} = e^{-sT} \frac{\omega_{n_a}^2}{s^2 + 2\zeta\omega_{n_a}s + \omega_{n_a}^2} \times \frac{\omega_{n_b}^2}{s^2 + 2\zeta\omega_{n_b}s + \omega_{n_b}^2}$$

$$\times \frac{\omega_{n_c}^2}{s^2 + 2\zeta\omega_{n_c}s + \omega_{n_c}^2} \times \frac{\omega_{n_d}^2}{s^2 + 2\zeta\omega_{n_d}s + \omega_{n_d}^2}$$

$$\times \frac{\omega_{n_e}^2}{s^2 + 2\zeta\omega_{n_e}s + \omega_{n_e}^2}. \qquad (4.18)$$

The overall performance of this department is illustrated in Figure 4.14. The delay factor results in zero output until time T. The objective of an organization is to reduce this factor to as small a value as possible.

Utilizing mathematical models of the organization, the systems engineer can proceed to apply conventional control theory to the schedule

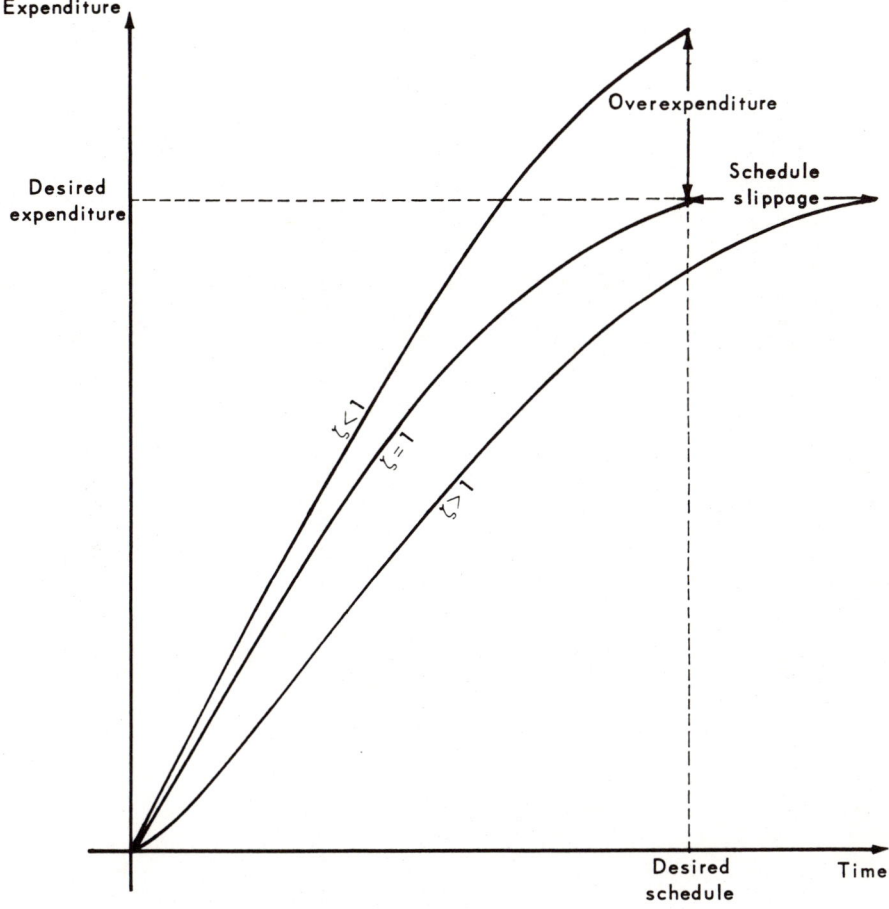

Figure 4.12. Underdamped, Overdamped, and Critically Damped Organizational Responses

and cost problem. This is an extremely powerful potential tool for modern management control. An actual program's case history is analyzed in the following section of this chapter, utilizing this approach.

Program Performance—An Actual Case History

An actual example of an organization's dynamic performance is presented in this section. It is based on information presented in Reference 3. The reasons for studying this case history are twofold: It provides practice in

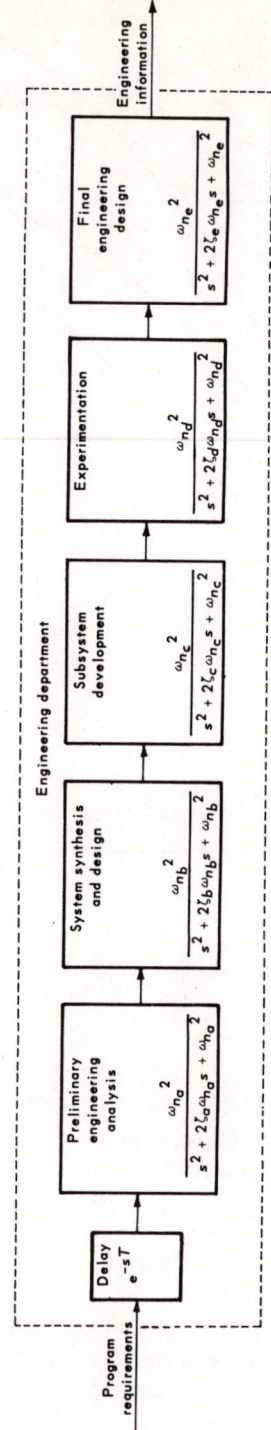

Figure 4.13. Mathematical Model Representation of an Engineering Department

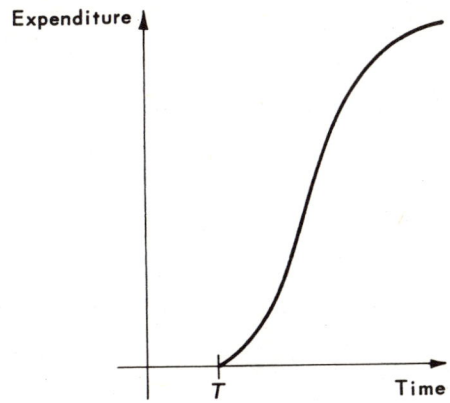

Figure 4.14. Example of an Engineering Department's Response

applying the principles previously presented for determining the mathematical model of an organization's performance, and provides an understanding of what can be done to improve its performance.

Let us assume that the engineering organization under study is composed of three departments: Engineering, Drafting, and Fabrication. It is producing an electronic device to be used for a large system. The value of the contract is $30,000 and the delivery schedule is 24 months. Figure 4.15 illustrates the theoretical and actual expenditures over the period of the contract.[3] In addition to the expenditures of each of the three departments, the total accumulated costs are also shown. A brief overall analysis indicates that the program was completed 4 months earlier than scheduled and underran the budget by about $800.

Let us next apply the tools previously presented to synthesize a mathematical model of the organization's performance. This is very useful for understanding the organization's dynamic characteristics and capability for future programs. The various transfer functions are indicated in Figure 4.15. Preliminary engineering occurs during the first 4.6 months of the program and behaves as a simple first order lag network whose transfer function is given by

$$Y_1(s) = \frac{1}{1.5s + 1}. \qquad (4.19)$$

The time constant of the lag network is 1.5 months. The design and experimentation engineering phase does not start until 4.6 months after the start of the program and also behaves as a lag network whose time constant

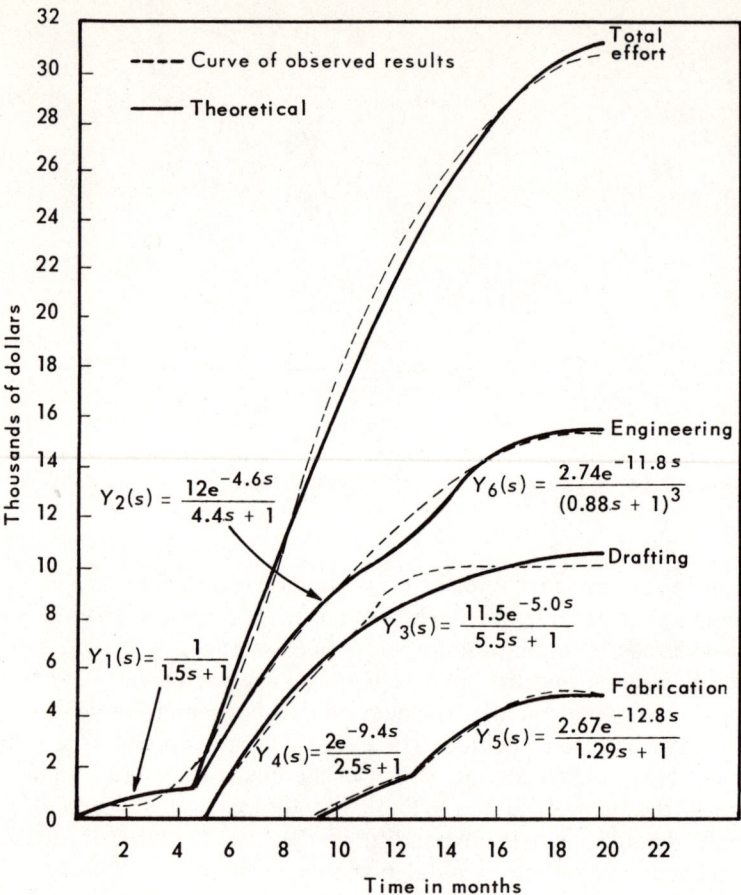

Source: Wilcox, Robert B., "Analysis and Synthesis of Dynamic Performance of Industrial Organizations—The Application of Feedback Control Techniques to Organizational Systems," *IRE Transactions on Automatic Control* AC-7, March 1962, page 65. Reprinted with permission.

Figure 4.15. Mathematical Models of Cost Curves

is 4.4 months. In addition, it has a transportation lag due to its delay in starting. This transfer function is given by

$$Y_2(s) = \frac{12}{4.4s + 1} e^{-4.6s}. \qquad (4.20)$$

Engineering test commences 11.8 months after the start of the program and is best represented as a triple lag network with a transportation lag:

$$Y_6(s) = \frac{2.74}{(0.88s + 1)^3} e^{-11.8s}. \qquad (4.21)$$

The drafting effort begins 5 months after the start of the program and can best be represented as a lag network with a transportation lag:

$$Y_3(s) = \frac{11.5}{5.5s + 1} e^{-5s}. \qquad (4.22)$$

The fabrication effort is composed of two segments: Production and Assembly. The production phase begins 9.4 months after the start of the program and continues for 3.4 months. Its transfer function can best be represented by a lag network and a transportation lag:

$$Y_4(s) = \frac{2}{2.5s + 1} e^{-9.4s}. \qquad (4.23)$$

Assembly starts 12.8 months after the start of the program. Its transfer function can be represented as a lag network and a transportation lag:

$$Y_5(s) = \frac{2.67}{1.29s + 1} e^{-12.8s}. \qquad (4.24)$$

The overall performance of this program team can be represented by the following transfer function:

$$\frac{C(s)}{R(s)} = \frac{31.91}{24.2s^2 + 9.9s + 1}. \qquad (4.25)$$

This results in an ideal, critically damped system whose natural frequency is 0.2 months and whose time constant is 4.95 months. These factors are measures of the organization's ability to adapt to new work and develop a product.

Figure 4.16 is a simplified block diagram of the organization's dynamic performance based on the mathematical analysis performed in Figure 4.15. The dotted lines represent control signals exerted by the program manager. This mathematical model is an excellent aid for future program planning. By focusing attention on areas of major lags and delays, corrective action can be initiated to improve organizational performance and accurately predict future delivery schedules.

Estimating Schedules and Costs

The program manager relies heavily on actual program histories and statistical information for estimating schedules and costs. An understanding of an organization's dynamic performance, through the use of mathematical

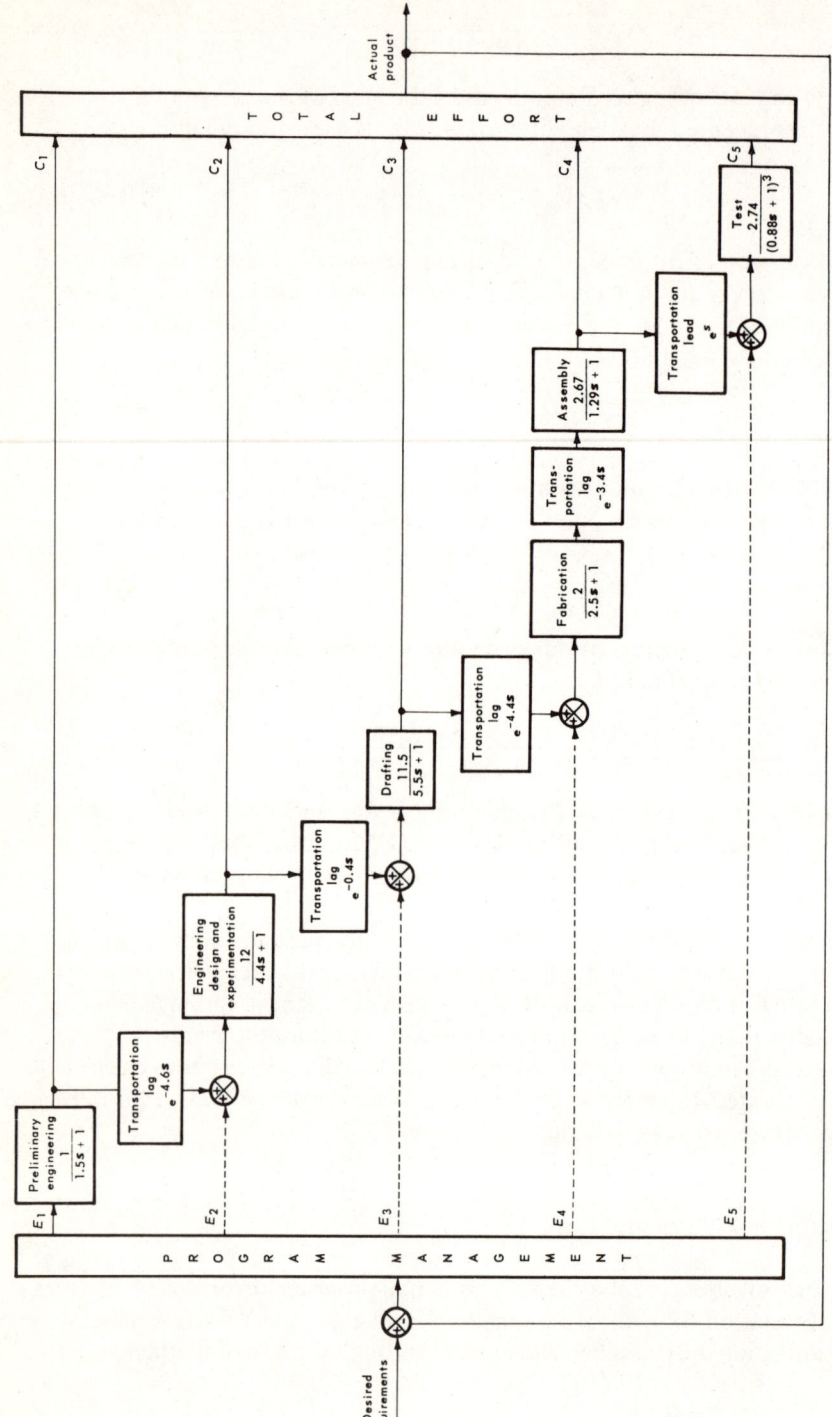

Figure 4.16. Block Diagram Representation of Program Illustrated in Figure 4.15

models, is very helpful for initiating improvements. The utilization of time models, as PERT, is very useful for controlling a program's schedule. In addition, such statistical information as the average cost per drawing, transistor, or other unit factor is helpful in estimating cost.

As discussed in the previous section, the overall performance-time relationship of an organization is similar to the transient behavior of a type 1, second-order control system. By synthesizing the transfer functions of the various components of the organization, the performance-time relationship of each element can be predicted and used to aid management in their decisions.

Analysis of an organization's mathematical models focuses attention on extreme delays and lags in performance. By reducing these delays and lags, greater use of facilities over a given period of time is possible. This has the effect of reducing the overhead by making greater utilization of available facilities and indirect support labor. Economically, this results in an increased return on assets which is a very key measure of successful operation and management of a business.

Utilization of time models, as PERT, is essential for predicting and controlling the schedule of modern, complex systems. As discussed previously in this chapter, this graphic tool displays all activities necessary to complete the program, provides information regarding the overall interrelationship of these activities, and can be used very effectively for controlling and reviewing the progress of a program. Its use, in all phases of schedule control, is strongly recommended.

The statistical correlation of such dependent variables as the man-hours needed for preliminary engineering, development engineering, test engineering, drafting, production, and assembly with such independent variables as the hardware pieces used in functions, circuits, and assemblies also provides a useful basis for estimating. The reader is referred to Martino[14] for procedures utilizing such a statistical approach. The establishment of a well-founded estimating system, based on statistics, is an extremely important aid for estimating costs of modern, large, complex systems.

In addition to modeling and statistical relationships, experience and knowledge of personnel are extremely important factors for estimating schedules and cost. The program manager must know the organization, its people, and capability. With this information at his disposal, realistic schedules and cost estimates can be prepared.

Use of Computers as Aids for Controlling Schedule and Cost

The digital computer is another very important tool used for predicting and controlling a program's schedule and cost. The accounting procedures of most organizations have been programmed on digital computers and most programs utilize the computer for controlling schedule and cost. Because of the computer's virtues of large capacity and high speed, the schedule and cost of large complex programs can be controlled and their status can be monitored instantaneously. In addition, computers can be used to estimate schedule and cost based on past statistical data and/or mathematical models of the organization.

A digital computer can make program control of large, complex systems meaningful. In addition to its great storage capacity and high speed of response, the computer also can be instructed to detect any event that does not conform to an expected pattern and to feed this information back to the program manager. Thus, time models such as PERT and mathematicals models of cumulative performance can be practically implemented.

As an example, consider the implementation of PERT for a large program utilizing a computer. The digital computer plays the part of a staff of clerks and mathematicians by recording and performing various mathematical operations. Such mathematical characteristics as the standard deviation, variance, the statistically expected time, latest allowable time, and slack time can be computed.

Two basic types of inputs must be supplied to the computer. The first category includes information concerning the values of t_a, t_b, and t_c. In addition, schedule dates, completion dates, and appropriate codes which indicate the type of operation to be performed are inserted. The second category supplies information to the computer regarding the title of each event, its number, and operation codes for performing changes later.

The computer prints out a variety of information, generally including event number, event nomenclature, expected date, latest allowable date, schedule date, actual date, slack time, standard deviation, and probability. In addition, some computers print bar graphs from which the expected time and the latest allowable time can readily be determined for each event. With this tool, the system manager can make rapid assessments of any changes.

The biggest disadvantages of utilizing a digital computer in this phase of the operation are the relatively high costs associated with its initial purchase, programming, maintenance, and operation. Computers require a substantial investment and highly specialized skills to operate. As with any other piece of equipment, it attains its full value only when used to the fullest extent possible. This brings up the question of whether every program's schedule and cost should be controlled by a computer. The management of each business must ascertain at what threshold level it desires to monitor a program's status via computer. For example, let us assume that organization ABC, with annual sales of 100 million dollars, receives a contract award of 10 million dollars to produce a complex weapons control system composed of 89 subsystems. The only logical manner to control the schedule and cost of this program smoothly is to utilize a computer, together with such powerful tools as time models (PERT) and perhaps mathematical models of the organization's performance. As a second example, consider the XYZ company. It does an annual business of one million dollars and receives a contract award for $16,000 to produce a control console for a hydroelectric power station. In most cases, a company of XYZ's size does not have its own computer facility, and must rent time from a centrally located unit. The cost of programming this project on a computer, assuming that one is available, is prohibitive compared to the size of the award. As a third example, let us assume that the ABC company receives the award of $16,000. Should it control such a relatively small program via a computer? In companies with a centrally located, large-scale computer which has standardized the schedule and cost control programming, it may be economically feasible. The dividing line of where a program is too small to be controlled by a computer, for a business the size of ABC, is not very clear. Good judgment by management is needed in this area.

References

1. Campbell, D.P. "Dynamic Behavior of Linear Prediction Systems." *Mechanical Engineering* (April 1953).
2. Forrester, J.W. *Industrial Dynamics*, Cambridge, Mass.; MIT Press, 1961.
3. Wilcox, Robert B. "Analysis and Synthesis of Dynamic Performance of Industrial Organizations—The Application of Feedback Control Techniques to Organizational Systems." *IRE Transactions on Automatic Control* AC-3, (March 1962).
4. Gaddis, P.O. "The Project Manager," *Harvard Business Review*, May-June (1959).
5. Frantz, Robert A. Jr. and Northern, Lloyd B. "Project Control Using PERT," *Electro-Technology* (August 1964).
6. Reres, Joseph A. "PERT—Here's How It Works," *E/E Production*, (September-October 1962).
7. *DOD and NASA Guide–PERT Cost System Guide*, Office of the Secretary of Defense and Office of the Administrator, National Aeronautics and Space Administration, Washington, D.C., June 1962.
8. *NASA, PERT and Companion Cost System Handbook*, Director of Management Reports, Office of Programs, National Aeronautics and Space Administration, Washington, D.C., October 1962.
9. Mason, S.J. "Feedback Theory: Some Properties of Signal Flow Graphs," *Proc. IRE* 41 (Sept. 1953) p. 1144.

10. Mason, S.J. "Feedback Theory: For the Properties of Signal Flow Graphs," *Proc. IRE* 44 (July 1956) p. 920.
11. *PERT–Summary Report, Phase 1*, Special Projects Office, Bureau of Naval Weapons, Dept. of the Navy, Washington, D.C., July 1958.
12. Wolf, E.A., *Program Evaluation Review Technique*, Brooklyn: Polytechnic Institute of Brooklyn, 1963.
13. Chestnut, H. and Mayer, R.W. *Servomechanisms and Regulating System Design*, vol. 1. New York; John Wiley & Sons, Inc., 1959.
14. Martino, P.V. "Predicting Product Development Costs," *Machine Design* (February 21, 1953).

5 Simulation

Introduction

Simulation is a very important technique widely used in many phases of systems engineering. The process involves the generation of a simplified mathematical model of the system in order to represent the system's characteristics. Simulation does not include all the features and characteristics of the original system. Instead, it focuses attention on certain factors of the system which are of interest.

Simulation offers the following advantages:[11]

1. It is a convenient method of quickly evaluating alternative preliminary systems configurations to determine their anticipated performance.

2. The analysis of interactions among subsystems can be determined.

3. It is a convenient method of modeling situations which can't be easily determined from the actual physical system.

4. Modes, inputs, situations, and configurations which cause system instability or malfunctions can be determined.

The descriptive mathematical models used in systems simulation represent a set of formal statements concerning an ideally hypothesized operation. Although models may not represent an actual system in all respects, they do describe its essential inputs, characteristics, and outputs. In addition, it provides an indication of environmental conditions similar to those found in the actual system. For example, a block diagram of a control system containing transfer functions can be considered a mathematical model, and it permits relationships to be determined between the input and output of the system. PERT, discussed in Chapter 4, is an example of a time model which can be put on a computer simulation for determining schedule, and for studying the causes and effects of certain schedule variations.

Systems simulations can be synthesized using *deterministic* or *stochastic models*. *Deterministic models* can be used when an input to the system always results in a corresponding output. Control systems block diagrams containing transfer functions, analyzed in Chapter 2, are deterministic models. *Stochastic models* are used to represent systems in which the system makes a random choice from a set of permissible systems parameters before generating an output response to an input. By their very characteristics, stochastic models yield only average responses. To obtain confidence in the results obtained from stochastic models, a large number of iterative trials are required to generate a large statistical sample size. Obviously stochastic models are much more complex than deterministic models and are used in practice only when absolutely necessary.

Another method of categorizing systems simulation is by performing mathematical computations. For example, it is common to refer to the simulation by naming the type of computer equipment being utilized, such as analog, digital, or hybrid.

Chronologically, analog computers came first and are still very useful for solving certain types of system problems. The digital computers became relatively accessible in the late 1950s and have grown in importance and usage ever since. The use of digital computers for simulation is much greater than that of analog computers today. Hybrid computers, which are a combination of analog and digital techniques, have also recently become important for systems simulation.

The choice of the type of simulation to be used, deterministic or stochastic, and analog or digital or hybrid is a problem of systems optimization. The selection is usually based on considerations of the availability of certain types of computers, schedule, cost, and the types of answers desired.

Analog Simulation[9]

The process of analog simulation utilizes a continuous model which represents the mathematical characteristic of a physical system. Electrical analogs are used to represent the mathematical operations of the model.

The analog computer performs computations, memorizes, and makes logical decisions on variables which appear in continuous form. The classical analog computer simulation problem is to simulate a system which may have several unknown parameters. By observing the system's transient response, the various unknown parameters can be adjusted until the desired performance is achieved.

The system need not be limited to purely linear elements. In addition, the control system simulated need not be limited to systems whose differential equation is of low enough order to be handled by conventional mathematical techniques. Therefore, the analog computer can be used to simulate complex systems containing nonlinear characteristics such as saturation, hysteresis, dead zones, etc.

The analog computer is very well suited to integrate. This feature is very important from the viewpoint of control systems which are very dependent on integration. In addition, the analog computer is limited only by the bandwidth of its respective amplifiers.

Perhaps the biggest disadvantage of analog computers are their accuracy. The machine variables are usually electrical voltages that range from about +100 to −100 volts. Under ideal laboratory conditions, the accuracy can approach 0.01 percent of full scale. However, 0.1 to 1.0 percent is much more common for the typical analog computers commercially available.

The analog computer has a poor memory capability. Machine variables can be memorized only for the period of time that the integrator's capacitors can retain them.

The analog computer simulation also presents the problem of "scaling." This is concerned with the proper setting of attenuator and amplifier gains without causing saturation of the system and limiting its noise.

Analog Computer Elements[9]

Analog computers are relatively simple to visualize due to the ease with which the fundamental operations of multiplication by constants, addition, differentiation, and integration can be performed. Multiplication of variables, and generation of arbitrary functions can also be performed relatively simply. This section reviews some of the basic analog computer elements.

The fundamental component of an analog computer is a high gain d.c. voltage amplifier. By appropriately choosing the input and feedback impedances, any desired analog computer characteristic can be obtained easily. These devices are of relatively high gain in the range of 10^5 to 10^7. Due to these high gains, the drift problems associated with d.c. amplifiers are greatly magnified. To minimize the drift, highly regulated power supplies, temperature-compensated precision resistors, and feedback techniques are designed into each d.c. amplifier.

Let us consider the basic operational amplifier circuit illustrated in Figure 5.1. It consists of an input circuit impedance, Z_i, a feedback impedance, Z_f, and the input impedance of the d.c. amplifier, Z_g. The open loop gain of the amplifier is assumed to be K. The gain of the feedback circuit, e_o/e_i, can be obtained from the following set of equations:

$$e_i - e_g = i_i Z_i \tag{5.1}$$

$$e_o - e_g = i_f Z_f \tag{5.2}$$

$$e_g = (i_i + i_f) Z_g \tag{5.3}$$

$$e_o = -K e_g. \tag{5.4}$$

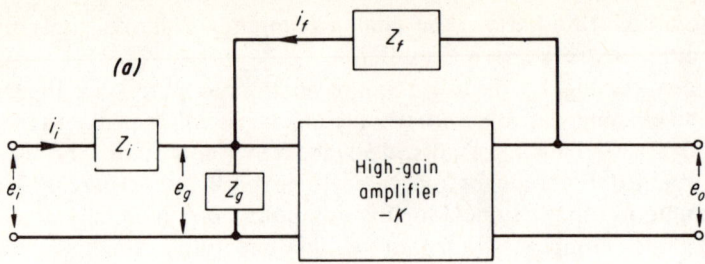

Figure 5.1. Basic Operational Amplifier Circuit

Solving Equation (5.1) and (5.2) for i_i and i_f, respectively, and then substituting these values into Equation (5.3), the following equation is obtained:

$$e_g = \frac{e_i - e_g}{Z_i} + \frac{e_o - e_g}{Z_f} Z_g . \tag{5.5}$$

Solving Equation (5.5) for e_g and substituting it into Equation (5.4), the overall gain of the feedback operational amplifier is obtained.

$$\frac{e_o}{e_i} = -\frac{Z_f}{Z_i} \frac{1}{1 + \frac{1}{K}\left[1 + \frac{Z_f(Z_i + Z_g)}{Z_i Z_g}\right]} . \tag{5.6}$$

In practice, Z_g is usually very much larger than Z_i and Z_f. In addition, as mentioned previously, K is a very large number, greater than 10^5. Therefore, Equation (5.6) can be approximated by the following relationship:

$$\frac{e_o}{e_i} \approx -\frac{Z_f}{Z_i} . \tag{5.7}$$

The corresponding simplified representation of the operational amplifier is illustrated in Figure 5.2.

Observe from Equation (5.7) that the ratio of Z_f to Z_i determines the characteristics of the operational amplifier. For example, if Z_i and Z_f are resistances of equal value, we obtain a sign changer. However, if Z_f is greater than Z_i, we obtain an amplifier. In addition, if the feedback is a capacitor and the input element is a resistor, the operational amplifier behaves as an integrator where

$$e_o(t) = e_o(0) - \int_0^t \frac{1}{RC} e_i(t) dt . \tag{5.8}$$

By reversing the capacitor and resistor, a differentiator results where

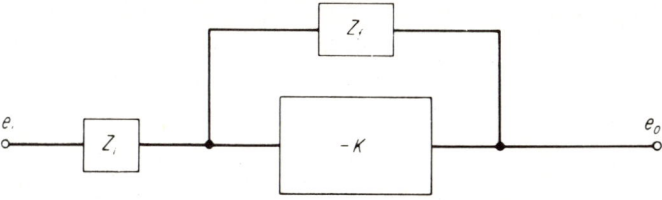

Figure 5.2. Simplified Representation of the Operational Amplifier

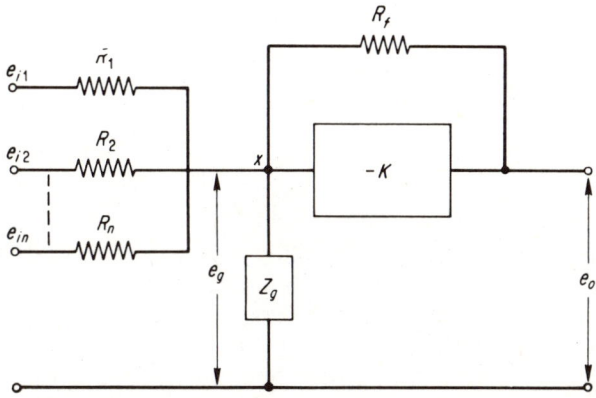

Figure 5.3. An Analog Adder

$$e_o(t) = RC\frac{d}{dt}e_i(t). \quad (5.9)$$

The circuit of Figure 5.1 can be easily modified to produce an analog adder, as indicated in Figure 5.3. This circuit can be analyzed easily if it is recognized that e_g is essentially zero for operational amplifiers. This can be proved for the circuit of Figure 5.1 from Equations (5.1) through (5.4). Solving for the ratio of e_g to e_i, and simplifying, the following equation is obtained:

$$\frac{e_g}{e_i} = \frac{1}{1 + \frac{Z_i}{Z_g} + \frac{Z_i}{Z_f} + K\left(\frac{Z_i}{Z_f}\right)} . \quad (5.10)$$

Since K is greater than 10^5, this equation clearly illustrates that e_g is maintained practically at zero volts. This is also true for the operational amplifier circuit of Figure 5.3. Therefore, by summing all of the currents flowing into node X of Figure 5.3 the following equation is obtained.

$$\frac{e_{i_1} - e_g}{R_1} + \frac{e_{i_2} - e_g}{R_2} + \ldots + \frac{e_{i_n} - e_g}{R_n} + \frac{e_o - e_g}{R_f} = \frac{e_g}{Z_g}. \quad (5.11)$$

Since we proved that e_g is zero, this equation reduces to the following form:

$$e_o = -\frac{R_f}{R_1} e_{i_1} - \frac{R_f}{R_2} e_{i_2} - \ldots - \frac{R_f}{R_n} e_{i_n}. \quad (5.12)$$

Equation (5.12) can be put in the following simpler format:

$$e_o = -\sum_{n=1}^{N} A_n e_{i_n} \quad (5.13)$$

where

$$A_n = R_f/R_n.$$

Equation (5.13) clearly illustrates the additive characteristic of this circuit. The analog gains associated with each input are conventionally marked on the block diagram of the operational amplifier.

In the case of operational amplifiers, one or more components of the device are purposely left out in order to make it more flexible. These components can then be externally selected in order to vary the closed-loop gain and bandwidth, and have the operational amplifier perform as an integrator, sign changer, or adder.

Integrated circuits techniques have revolutionized the operational amplifier. It has all the desirable attributes one was striving to achieve while first using vacuum tubes, and then transistorized operational amplifiers: increased reliability, reduced size, and reduced power.

Table 5.1 summarizes the operation, circuit, and symbology of several common analog computing devices. The multiplier of two variables, shown as Operation 2, depends on a servo which positions the shaft of a potentiometer.[1] The circuits associated with Operations 3 through 6 depend on the high gain of a d.c. amplifier, K. The higher the gain of this amplifier, the more exact is the operation performed. The function generator shown in Operation 7 is usually performed by means of nonlinear elements such as diodes.[2] By properly interconnecting and biasing them, these diode circuits can generate any desired nonlinear function.

As mentioned previously, an electronic differentiator can be constructured by reversing the capacitor and resistor of the circuit shown in Operation 6 of Table 5.1. However, systems engineers do not use differentiators

Table 5.1
Summary of Basic Analog Computing Elements

Operation	Circuit	Symbol
Multiplication of a variable by a constant: $e_o = a e_i$ where $0 < a < 1$		
Multiplication of two variables: $e_o = a e_1 e_2$ where $0 < a < 1$		
Multiplication and sign change: $e_o = -A e_i$ where $A = R_f/R_i$		
Addition (case 1): $e_o = -\sum_{n=1}^{N} A_n e_{in}$ where $A_n = R_f/R_n$		
Addition (case 2): $e_o = -\sum_{n=1}^{N} A\, e_{in}$ where $A = R_f/R$		
Integration: $e_o(t) = e_o(0) - \dfrac{1}{RC}\displaystyle\int_0^t e_i(t)\,dt$		
Function Generation: $e_o = f(e_{i1}, e_{i2}, \ldots, e_{in})$		

in practice since they have a tendency to amplify noise due to their wide bandwidths. In addition, their very good high frequency response may lead to unwanted high-frequency oscillations in the computing machine unless special precautions are taken. Actually, by arranging the elements properly, it is possible to use integrators instead of differentiators for simulation on an analog computer. This will be illustrated with several examples in this chapter.

Analog Computer Simulation Techniques[9]

The purpose of this section is to illustrate how the systems engineer may simulate a system on an analog computer directly from the system's differential equations. The problems considered here are elementary. However, the basic techniques presented are applicable to all kinds of complex problems.

Consider the simulation of a system that can be represented by the following linear differential equation model having constant coefficients:

$$\frac{d^2c}{dt^2} + B\frac{dc}{dt} + c = F. \tag{5.14}$$

Further assume that the initial conditions are zero, and that $B > 1$. Solving for the highest order derivative, we obtain

$$\ddot{c} = -B\dot{c} - c + F. \tag{5.15}$$

This equation states that the analog voltage $\ddot{c}$ must appear as a sum of the analog voltages $B\dot{c}$ and c as illustrated in Figure 5.4.

The voltages $B\dot{c}$ and c, which must appear at the input to the summing amplifier are not independent of $\ddot{c}$, but must be obtained from $\ddot{c}$ through successive integrations. This can be performed by assuming that the voltage $\ddot{c}$ is already known and applying it to the input of an integrator whose output is $-\dot{c}$. Successive integration then yields the output c. Figure 5.5 illustrates the analog computer block diagram corresponding to the given differential Equation (5.14). Observe that the system's transient response for a step input can be obtained by applying the input to another input of the adder. The output can be recorded from c, and B can be adjusted to achieve the desired characteristics.

As a second example, consider the simulation of a system whose model consists of two differential equations. These equations must be solved simultaneously on an analog computer. Assume that each differential equation is a function of two dependent variables x and y as follows:

$$\frac{d^2x}{dt^2} + \frac{dx}{dt} + x + 2y = F \tag{5.16}$$

$$\frac{d^2y}{dt^2} + 4\frac{dy}{dt} + 2y + 8x = 0 \tag{5.17}$$

In addition, assume that the initial conditions are zero. The procedure in this example is similar to the preceding example. The assumption is made that the highest order derivative of each differential equation is known, and lower order derivatives are solved by means of successive integrations and multiplications by constants. Therefore, two analog computer block di-

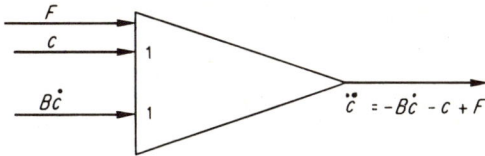

Figure 5.4. Relationship Between c, $B\dot{c}$, and $\ddot{c}$

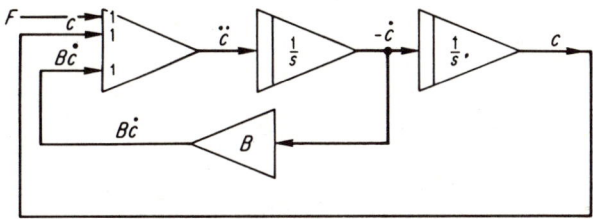

Figure 5.5. Analog Computer Block Diagram Corresponding to Solution of the Differential Equation: $\ddot{c} + B\dot{c} + c = F$

agrams are obtained which are dependent on each other. Figure 5.6 illustrates the overall analog computer block diagram. The outputs x and y can be recorded and will vary in time in accordance with Equations (5.16) and (5.17).

There is usually more than one correct analog computer simulation block diagram for any given problem. In general, we attempt to minimize the number of computing elements and attempt to keep the gains of the d.c. amplifiers as low as possible due to drift and noise considerations. In other specific problems, special considerations may dictate other considerations.

These two illustrative simulation problems have assumed that the initial condition of the system was zero. In general, it is not, and it is necessary to establish at the output of each integrator in the analog computer a voltage equal to the given initial value of the variable which is to appear at that point. Since the integrator's memory is relatively short due to capacitor leakage problems, the initial conditions cannot be maintained constant before each run while the analog computer is set up to solve particular differential equations. Therefore, the analog computer is made inoperative by disconnecting the integrator's feedback network from the input to the amplifier. This is conventionally denoted as the RESET condition of the

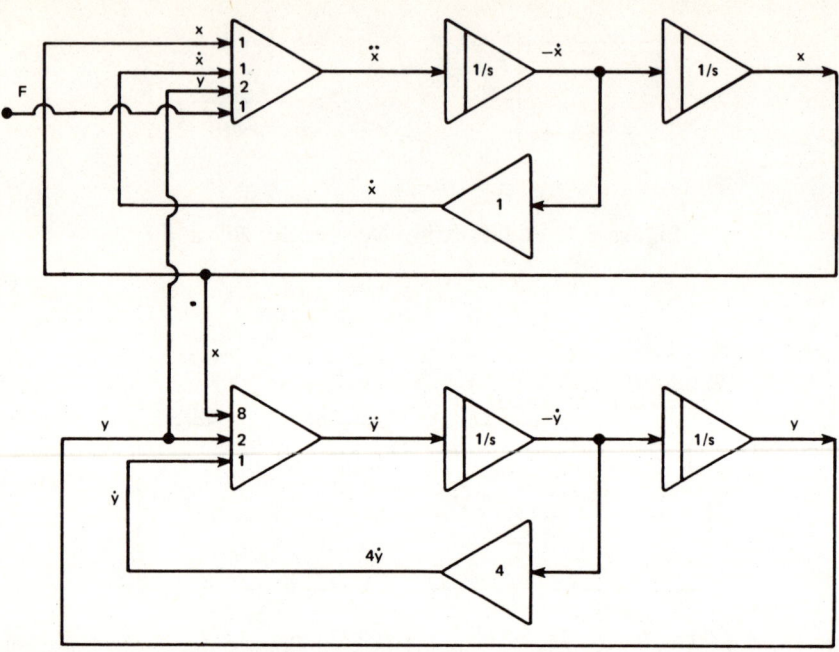

Figure 5.6. Analog Computer Block Diagram for Solving the Simultaneous Equations: $\ddot{x} + \dot{x} + x + 2y = F; \ddot{y} + 4\dot{y} + 2y + 8x = 0$

analog computer. At the beginning of a computer run, a relay system connects the amplifier to the feedback network and at the same time removes the initial condition source voltages. Then the integrating capacitors are charged up to the correct voltages and the analog computer variables are free to vary according to the differential equations' requirements. Summing amplifiers and inverters need not be made inoperative in the RESET condition. Their output voltages will assume the correct initial values of the voltages established at the amplifier input terminals from preceding integrators.

The actual solution time of a differential equation may sometimes be too fast for the recorder to follow its response accurately. In addition, there are cases where the overall solution may take a very long time. These problems can be overcome by changing from real time to a synthetic "machine time" for use by the analog computer. This can be accomplished easily by substituting $t = \tau/a$ in the differential equation. If $a < 1$, then the analog computer solution is "speeded up." Alternately, if $a > 1$, the solution is "slowed down."

Digital Simulation[9]

Digital simulation of large, complex systems is widely used today. The impetus for the great trend toward the digital computer simulation is the ready availability of large computer centers.

Digital computers are capable of performing long sequences of computations without human intervention. They can also make certain logical decisions and even change their future actions as a result of these decisions. The digital computer simulator basically performs the computation, memorization, and logical decisions in terms of the model variables which always appear in discrete form as data sequences.

The ability of digital computers to make precise calculations and decisions at very high speeds has made it possible to use them in real-time systems simulations. Systems which use digital computers as part of the controlling action are called "real-time discrete systems" because information must be processed and discrete decisions must be made in real time. Examples of these are digital computers which control automatic processes in the chemical industry, those which are used in machine tool guidance and control, and those used aboard space vehicles to perform the necessary attitude control functions.

In comparison to analog computer simulations, digital computer simulations have several advantages: There is no actual limit to the accuracy obtainable if we have sufficient computing time; the digital computer is very reliable; stored information can be kept indefinitely; there is no scaling problem. The major relative disadvantages are high cost and very poor man-machine communications. In addition, we cannot perform true integration since we are dealing with difference rather than differential equations. This is usually handled by utilizing an interpolation method on the discrete steps taken.

A digital computer can be used directly in the control loop or it can serve outside the loop. The former case is denoted as "on-line computer"; the latter as "off-line computer control." Figures 5.7 and 5.8 illustrate the two methods. "Signal converters" are used to convert information between analog and digital formats. They are also popularly referred to as analog-to-digital and digital-to-analog converters.

The on-line digital computer system is utilized where the process must be controlled on a continuous basis. Since these computers are generally less available for time-sharing other duties, they are usually considered to be "special-purpose digital computers." The "off-line digital computer" system is suitable when the process changes very slowly. Based on laboratory analysis and other information known to the operator, the digital computer is requested to make regular computations for achieving the desired output from the system. Since digital computers utilized in this

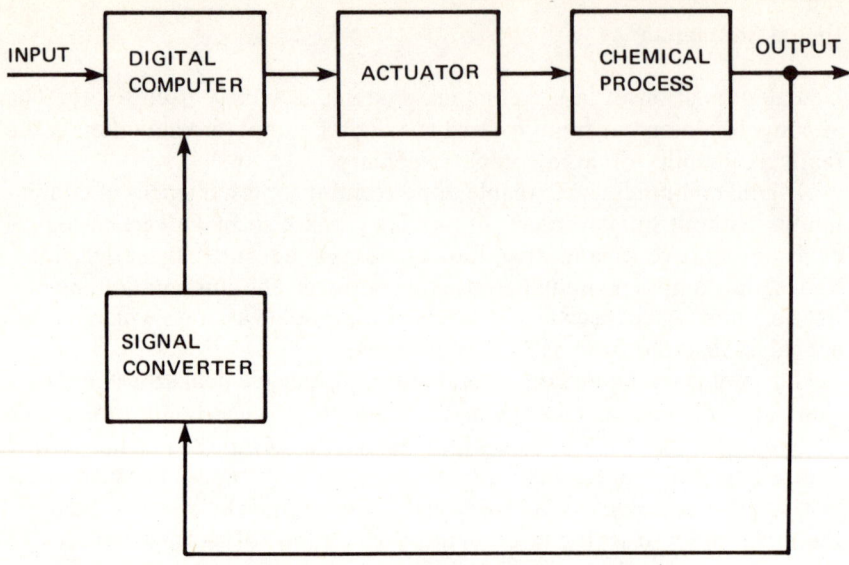

Figure 5.7. On-line Computer Control

manner will be available to perform many other tasks on a time-shared basis, we generally consider them to be "general-purpose digital computers."

Elements of Digital Computers[9]

The basic elements of a digital computer can be divided into 5 areas: input devices, output devices, control element, arithmetic elements, and storage.[3] Figure 5.9 illustrates the block diagram of a typical digital computer. Although commercially available machines vary considerably in details of the various components, the overall system concepts are basically the same.

The *input devices* read the input data into the digital computer. In addition, the instructions which constitute the program must also be read into the machine in most general-purpose computers. Examples of input devices are magnetic tape readers, punched-card readers, punched paper-tape readers, and various types of analog-to-digital converters.[4]

The *output devices* are used to record the computer's results. These are usually controlled by the control element. Examples of output devices are magnetic-tape machines, card-punching machines, oscilloscopes, special electromechanical typewriters, and high-speed printing devices.

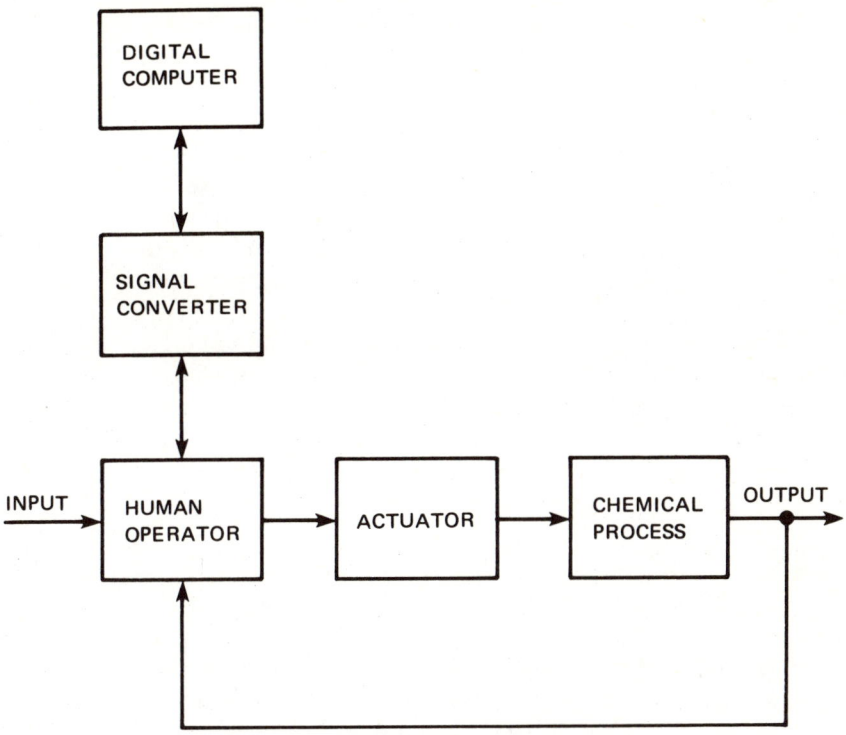

Figure 5.8. Off-line Computer Control

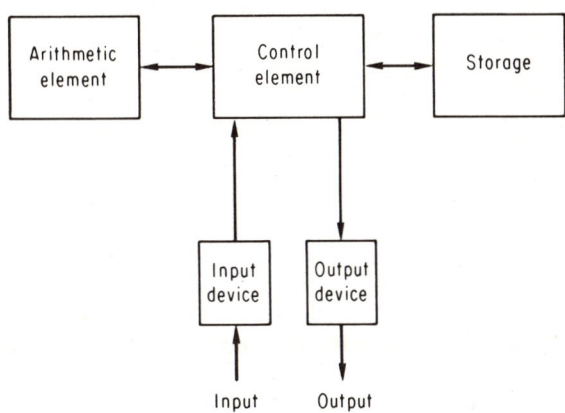

Figure 5.9. Block Diagram of a Typical Digital Computer

The *control element* of the digital computer sequences the operation of the machine and controls the actions of all other portions of the computer. The control circuitry interprets the instructions of the program and then orders the digital computer to perform the necessary operations.

The *arithmetic element* has the capability of performing addition, subtraction, multiplication, division, in addition to some logical operations. In many respects, the arithmetic element is similar to an ordinary calculating machine. The major difference is that the control unit tells the arithmetic element which of these operations it is to perform and supplies the necessary data for it.

The *storage* (or memory) section of the digital computer stores the information used during the computations. It is customary to divide the storage into segments denoted as an "address" or "memory location." Commonly used storage devices are punched cards, paper tape, magnetic tape, magnetic cores, magnetic drums, and magnetic discs.

Programming[3,9]

The man-machine interface with the digital computer is an extremely poor one. The computer needs information in discrete form and produces results in a like manner. The engineer utilizing a digital computer, of course, cannot easily communicate with such a machine. Therefore, it is necessary to write a program for the digital computer in a language it will understand.

The instruction words that direct the computer's operation are stored in it in numerical form. However, the programmer rarely writes his instructions in numerical form. Instead, each instruction to the computer is written using a letter code to designate the operation to be performed and the address in memory of the number to be used for this calculation. For example, an instruction word

ADD 466

means that the computer should add the number in store at address 466 as part of its calculation. Table 5.2 contains a summary of instruction words and their definitions commonly used.

It has been found that the majority of programs can be generated most readily by first drawing a logic flow chart which shows the computational steps required and then writing a step-by-step program. A *program* is basically a list of instructions which tells the digital computer how to perform its calculations. A method for accomplishing a particular set of calculations is denoted as an *algorithm*.

Let us consider developing the flow diagram for evaluating the following expression which could represent the simulation model of a system:

$$F = 3Y^3 + 2Y^2 + 4Y + 7. \tag{5.18}$$

Table 5.2
Definition of Instruction Words

Instruction Word		Function Performed by Instruction
Operation Code	Address Part	
CLA	430	The arithmetic element is emptied of all previous numbers and the number at address 430 is added into it. After the instruction is performed the arithmetic element contains the number in storage at address 430. CLA is a mnemonic code for "clear and add."
ADD	530	The number located at address 530 is added to any number which may be in the arithmetic element. After the instruction, the arithmetic element contains the sum of the number it previously contained and the number in address 530.
SUB	320	The number located at address 320 is subtracted from any number which may be in the arithmetic element. After the instruction, the arithmetic element contains the difference of the number it previously contained and the number in address 320.
STO	433	The number in the arithmetic element is stored at address 433. Any information previously in this address is destroyed. The number which was in the arithmetic element before the instruction was performed remains in the arithmetic element. This is generally referred to as a **STORE** instruction.
HLT	000	The machine is ordered to stop. The number in the arithmetic element remains.
MUL	400	The number at address 400 is multiplied by the number already in the arithmetic element.
DIV	500	The number already in the arithmetic element is divided by the number at location 500.
BRA	420	This instruction tells the computer to perform the instruction at address 420 next. The computer will skip or branch from the instruction it would have performed and perform instruction 420 instead. The computer will then perform instruction 421, followed by 422, etc.
BRM	420	The computer will branch to instruction 420 only if the number in the arithmetic element is negative. If the number is positive, the computer will not branch, but will perform the instruction stored at the next address in memory. **BRM** is a mnemonic code for "branch on minus."
PRT	200	Print out the number that is contained in address 200.

We want to calculate F for various values of Y. The required steps are illustrated in Figure 5.10. For this simple expression, the digital computer could be programmed to perform these calculations for several values of Y. In addition, it could perform the necessary computations several thousand times per second.

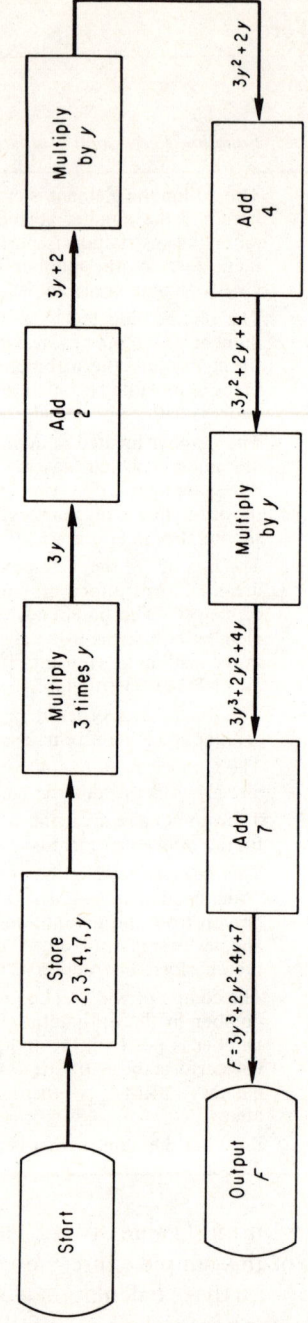

Figure 5.10. An Example of a Flow Diagram

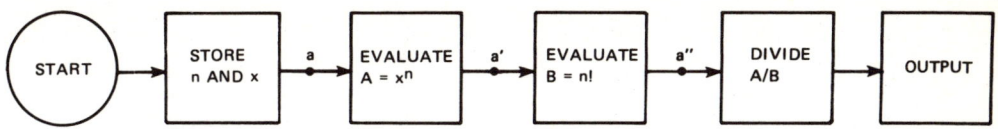

Figure 5.11. Flow Chart for Computing $y = x^n/n!$

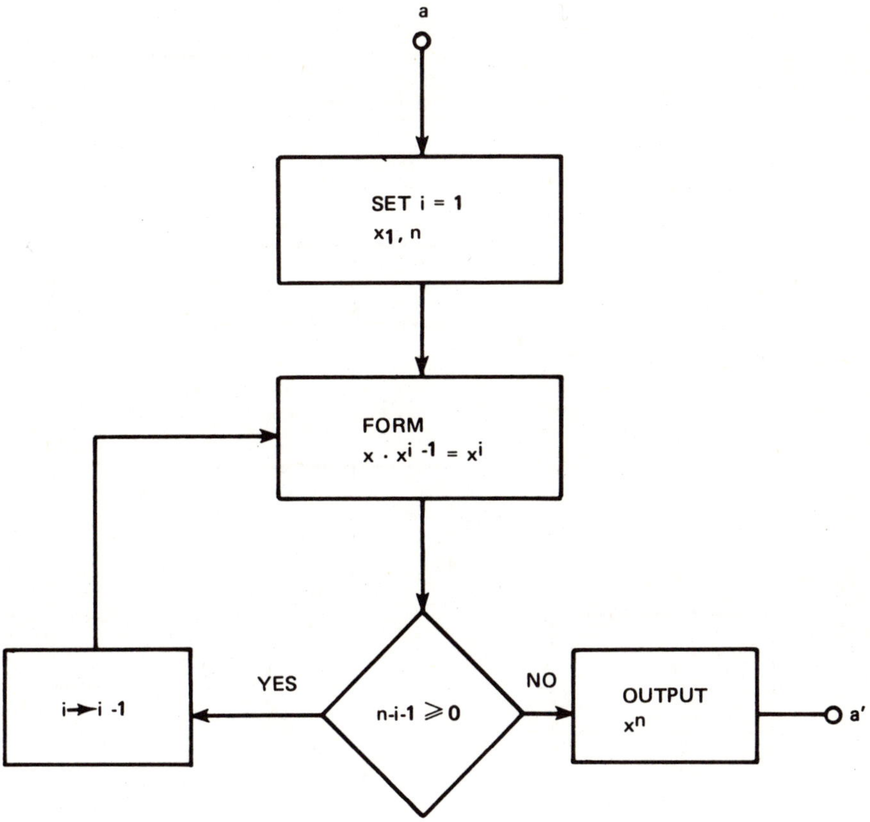

Figure 5.12. Detailed Flow Chart from Points a to a' of Figure 5.11

As a second example, let us consider the evaluation of

$$y = \frac{x^n}{n!} \tag{5.19}$$

which represents the simulation model of a system. A course flow chart is illustrated in Figure 5.11, and detailed segmented flow charts are shown in Figures 5.12 and 5.13. From these flow charts, the program shown in Table

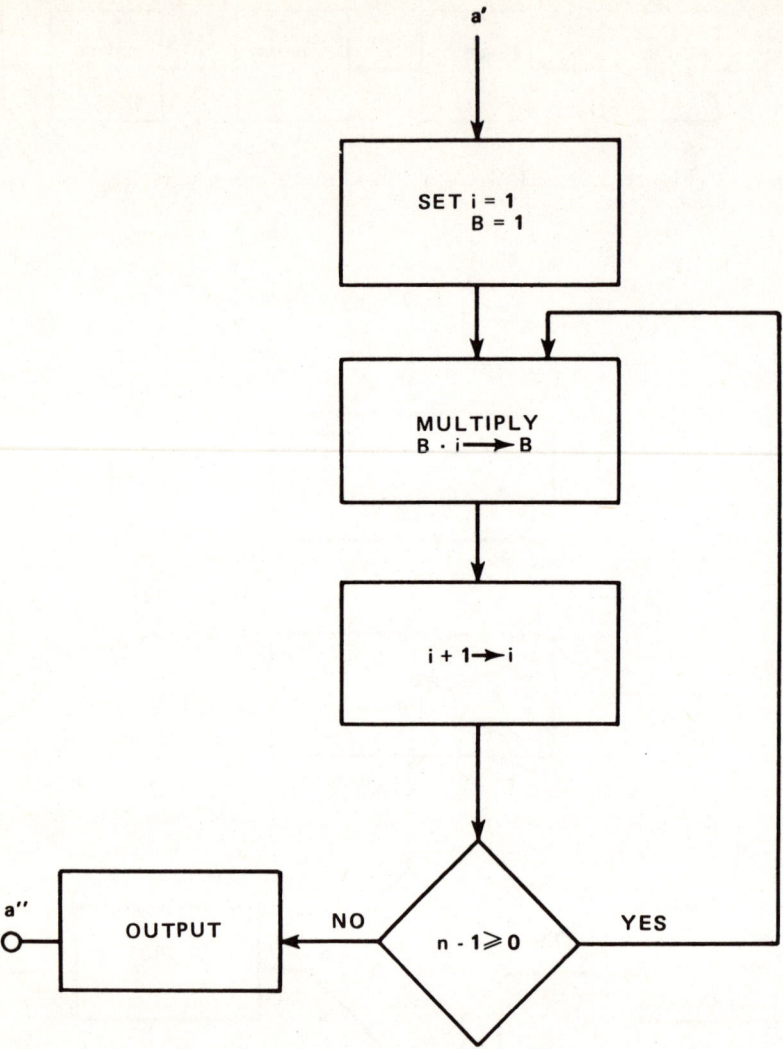

Figure 5.13. Detailed Flow Chart from Points a' to a'' of Figure 5.11

5.3 can be written. The reader should compare these figures and tables carefully to fully understand the steps involved in evaluating Equation (5.19) on a digital computer.

The digital computer is a great aid in programming. It is capable of translating written programs from a language which is straightforward and natural for the programmer into computer or machine language.

Table 5.3
Program for Determining $y = x^n/n!$

Store	1	x	n	Partial product cell	i cell
In Cell	100	101	102	104	106

Address	Operation	M	Description
	CLA	100	1 in Acc
	STO	104	
B	CLA	102	n in Acc
	SUB	100	$n-1$ in Acc
	STO	106	$n-1$ in i cell
	CLA	102	n in Acc
	SUB	106	$n-1$ in Acc
	STO	105	$n-1$ cell
	CLA	101	x in Acc
	DIV	105	$x/n-1$ in Acc
	MUL	104	Partial product in 104
	STO	104	
	CLA	106	i in Acc
	SUB	100	$i-1$ in Acc
	BRM	A	$i-1 \geq 0$
	PRT	104	
	HLT	000	
A	STO	106	
	BRA	B	

Therefore, programs are written to read other programs written in a language that was natural for the programmer, and then translating them into the computer's language. The program systems commonly used are denoted as *assemblers* and *compilers*. Each is a program that is intended to read a program written in a programming language and translating it into a computer language. The assembler or compiler is first read into the computer and is then followed by the program to be translated. After translation, the assembler or compiler stores the computer language program on punched cards, or paper or magnetic tape so that the program can be performed when desired. The language in which the programmer writes is called a *programming language*, and a program written in such language is called a *source program*. The translated program in computer language is called an *object program*.

The assembly program differs from the compiler program since its language closely resembles the computer's language. Each instruction to the digital computer in assembly language is generally translated into a single computer word. Therefore, an assembly language greatly resembles computer language. However, in compiler systems, a single instruction to the computer may be converted into many computer words.

The compiler language is the most advanced type of programming language, the simplest language to learn, and is used for most problems. However, compiler languages reveal very little about the computer on which they are run. As a matter of fact, many compiler languages are almost completely independent of the computer, and programs written in one of these languages may be run on any digital computer which has this type of compiler in its program library.

The most popular compiler languages are BASIC, FORTRAN, ALGOL, COBOL, FORMAC and JOVIAL. BASIC and FORTRAN are the most common computer languages used and several control programs using these languages are illustrated in the chapter problems. ALGOL was designed for international use and is used in describing algorithms for business purposes. COBOL is a language used for business systems. FORMAC does symbolic manipulation. JOVIAL is a language used for real-time systems. Obviously, there is no real universal language. A common feature of all of these computer languages is the *subroutine* which contains in package form all of the machine language instructions necessary to perform a specific computation. For example, if it is desired for the digital computer to find sin y, it can be commanded to do this by the FORTRAN statement **SIN (Y)**. The machine interprets this as a sequence of instructions that results in a computation of sin y.

Hybrid Computers[5,9]

Recently, there has been a growing trend toward hybrid simulation. The term hybrid computer refers to computers containing analog and digital segments. By combining a general-purpose analog and digital computer, the result is a hybrid computer containing the features of both facilities. For example, the resulting facility has the accuracy, speed, memory, decision, and arithmetic capability of the digital computer. In addition, it has the admirable man-machine interface and excellent integration capability in real time of the analog computer.

This is a relatively new kind of computer arrangement and its acceptance has been rather slow. Perhaps the biggest problem in its acceptance is the problem of having engineers understand both analog and digital computers. The usual problem is that most engineers prefer one or the other type of computer due to their lack of understanding of the other computer. It is my view, however, that the excellent features of the hybrid computer will force the issue in favor of their increasing use.

A typical general hybrid facility would consist of general-purpose analog and digital computers and linkage lines between the two. Since the two computers work with different types of signals, the main purpose of the linkage lines is to provide several digital-to-analog and analog-to-digital channels over which data information can be exchanged between the two

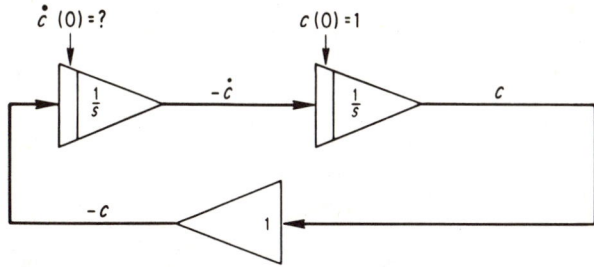

Figure 5.14. Analog Computer Simulation of the Split-Boundary Problem[5, 8]

computers. The functions of the linkage lines are for numerical and control purposes. The numerical data links send data between the two computers. The control links are used for controlling the behavior of the two computers. For example, signals from the analog computer have the ability to stop the calculations of the digital computer, and signals from the digital computer have the ability to control the mode of the analog computer's integrators. In addition, the potentiometers of the analog computer are automatically controlled by the digital computer.

The hybrid computer has a very high degree of flexibility. In addition to operating the digital and analog computers jointly, it can also be operated separately. I believe that the use of hybrid computers will increase greatly in the near future.

As an example of the type of simulation problem that a hybrid computer is very well suited for solving, we'll consider a problem presented in References 5, 8, and 9. It concerns a class of problems known as the *split-boundary* type in which the goal is to meet specified conditions at both ends of a control interval. For example, consider that the model of the system is given by the following undamped, second-order differential equation:

$$\ddot{c} + c = 0. \tag{5.20}$$

Instead of having specified the initial conditions of position and velocity, assume that the position at two values of time are specified as follows:

$$c(0) = 1,$$

$$c(2) = 1.$$

These equations state that the position is unity at $t = 0$ and 2 seconds later. Let us now try to simulate this system on an analog computer as illustrated on Figure 5.14. As indicated, we know only the initial condition for the second integrator, $c(0) = 1$, but not for the first integrator, $\dot{c}(0)$. To simulate Equation (5.20) on an analog computer, we need information concerning

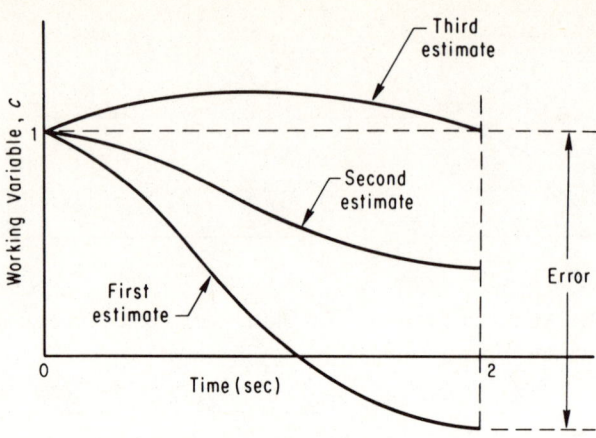

Figure 5.15. Effect of Successive Trials in Achieving $c(2) = 1$

the initial conditions of $c(0)$ and $\dot{c}(0)$. Therefore, how can we proceed, using an analog computer alone to determine $\dot{c}(0)$ which would make $c(2) = 1$?

For this simple problem, we could guess at a value for $\dot{c}(0)$, make an analog computer run, and check the result if $c(2) = 1$. If it doesn't work, another value of $\dot{c}(0)$ is chosen to reduce the error progressively as indicated in Figure 5.15. This figure illustrates that three successive runs were made until the correct value of $\dot{c}(0)$ was found.

Observe from this procedure that the simulation solution is iterative—the type of procedure that is very well suited for digital computers. Although in the case of the simple problem illustrated, it is relatively simple to keep changing $\dot{c}(0)$, in complex systems we would want these iterations to be performed automatically in a closed-loop manner by a digital computer. The digital computer would choose a series of values of $\dot{c}(0)$ and reduce the errors in a progressive manner until the error becomes zero. Therefore, a hybrid computer is ideally suited for this problem: the analog computer for simulating the analog portion; the digital computer to select the values of $\dot{c}(0)$ in an iterative manner, so that the error converges to zero and the desired value of $\dot{c}(0)$ is found. Figure 5.16 illustrates a conceptual arrangement of the automatic determination of $c(0)$ using a hybrid computer simulation.

Monte Carlo Methods[11,12,13,14]

The Monte Carlo method is a procedure used in stochastic simulations for obtaining approximate solutions of mathematical expressions which involve one or more probability distribution functions which may be inde-

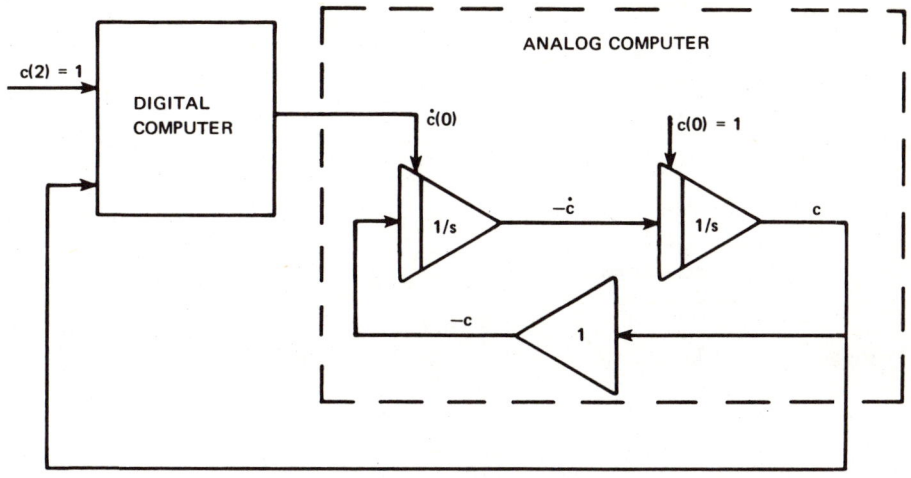

Figure 5.16. Automatic Hybrid Computer Determination of $\dot{c}(0)$

pendent of each other. Combining this technique with analytical methods, it is possible to obtain an average value and a dispersion for a solution to multiple probability distribution problems.

In a Monte Carlo problem, statistical results are obtained by repetitive sampling procedures. The accuracy of the Monte Carlo computations is a direct function of the number of samples used.

The basis of the Monte Carlo computation lies in the random selection of numbers from a given probability distribution. In practice, the method is used to generate random numbers with an equal likelihood of occurring. Each number has associated with it the resulting state that corresponds to this number.

To obtain an answer utilizing the Monte Carlo method, a particular situation must be simulated and run several times. The particular random numbers used are varied from run to run, but the same probability distribution is maintained. Computers are an excellent tool for performing these computations. The resulting answer is a random variable with a certain distribution. Figure 5.17 illustrates a block diagram of the Monte Carlo method.

Monte Carlo techniques are applied in stochastic system simulation problems in order to simulate systems operation in which a random decision can be made from among a set of possible alternatives. In the simulation process, a certain amount of uncertainty is introduced into the decision process so that the system's performance can be analyzed. By appropriately choosing the random signal, a fairly realistic system simulation can be achieved.

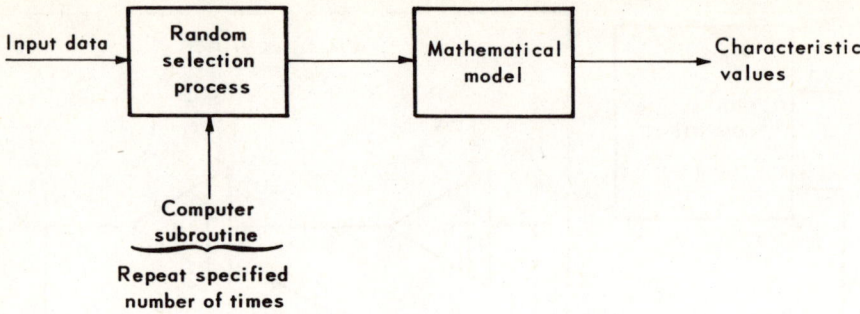

Figure 5.17. Block Diagram Representation of the Monte Carlo Method

As a simple example of applying Monte Carlo simulation techniques, consider the classical needle problem of Buffon.[16] A board is lined with a series of parallel and equidistant lines spaced at a distance h as illustrated in Figure 5.18. A needle, whose length l is shorter than the distance h, is thrown randomly on the board as illustrated. The probability that the needle will intersect one of the vertical lines is given by:

$$P = \frac{2l}{\pi h}. \tag{5.21}$$

Simulation of this experiment consists of assigning random values to the location of the needle midpoint x_m and the angle θ. Since the ruled lines are equidistant, it is necessary to consider only the interval $0 \leq x \leq h$.

The needle will intersect either of the vertical lines labeled $2h$ or $3h$ if

$$x_m < \frac{l}{2} \cos \theta \tag{5.22}$$

or

$$\frac{l}{2} \cos \theta > h - x_m. \tag{5.23}$$

If l is chosen to be 2, for example, then $(l/2) \cos \theta$ will vary between 0 and 1. Two random numbers q_1 and q_2 can be chosen from a uniform distribution of random numbers and it can be determined experimentally whether

$$q_1 > h - q_2 \tag{5.24}$$

and

$$q_1 > q_2. \tag{5.25}$$

The needle will intersect a line if either of these inequalities is valid;

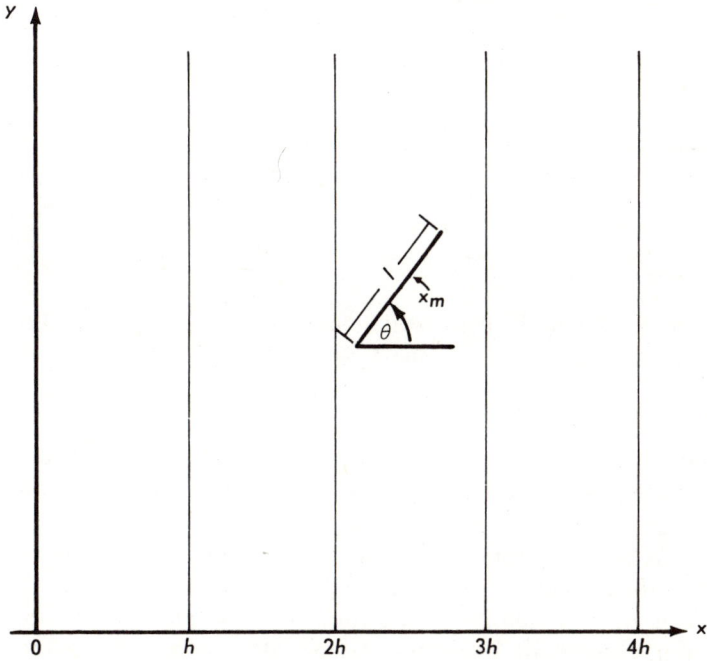

Figure 5.18. Buffon's Needle Problem

otherwise, there will be no intersections. By repeating this experiment N times, and X of these result in intersections, the probability of an intersection is given by

$$P_N = \frac{X}{N}. \qquad (5.26)$$

In the limit as $N \to \infty$, $P_N \to P$ as in Equation (5.21).

As a more practical simulation problem, consider the performance of a missile system. The ultimate performance of this type of system depends greatly on statistical factors such as random perturbations on the missile during operation. These consist of thrust variations, wind gusts, and tracking radar noise. In addition, possible evasive maneuvers on the part of the intended target must be considered. The performance of such a system is measured in terms of two probabilistic considerations: the "miss distance" of the missile, and the "probability of kill" for a given miss distance. Both of these factors are random variables which can be determined as statistical quantities.

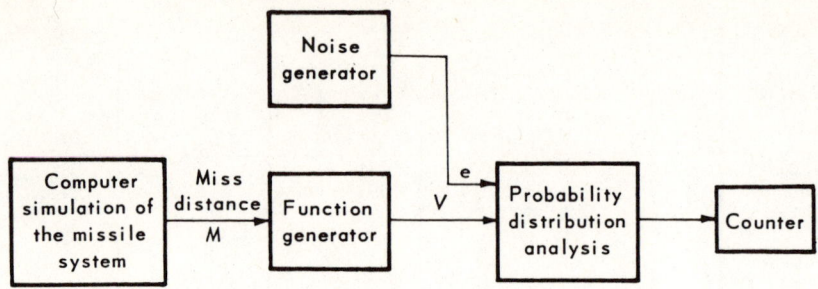

Figure 5.19. A One-Dimensional Solution for Determining the Probability of Kill of a Missile System

The probability of kill of the system in one dimension can be determined as illustrated in Figure 5.19.[15] Utilizing the computer, the missile system simulation contains appropriate statistical inputs which duplicate realistic conditions and generate an ensemble of miss functions. The statistical effect of the probability of kill for a given miss distance is evaluated by means of the probability distribution analysis and the random signal generator illustrated. The miss distance M controls the output of the function generator in such a way that the probability of the random signal generator output e being less than or equal to V corresponds to the probability of kill for every value of miss distance.

The probability distribution analysis then gates the output pulses to the computer with a probability of kill corresponding to the miss distance. The computer solves the problem repeatedly and generates a particular value for the miss distance on each run. The probability that the probability distribution analyzer will put out a pulse to be counted for each value of miss distance corresponds to the probability of kill for that particular value of miss distance. After several runs, the ratio of the number of counts to the number of runs is a direct measure of the probability of kill for the missile system.

The degree of error in the resulting simulation solution can be limited by the number of runs performed. The exact number of samples required can be estimated based on the degree of precision desired. The availability of flexible digital computers permits the use of Monte Carlo simulation techniques for investigating the overall effects of random errors having various distributions.

An important element of the Monte Carlo simulation is the random signal generator. Various methods for generating the random signal are considered in the following section.

Random Signal Generators

Random signal generators are required for stochastic systems simulation to introduce a controlled noise process whose statistical characteristics are known. These can be continuous for analog simulation or discrete for digital simulation. Random signal generators can be obtained from a variety of physical processes having an inherent random character.[16,17] This section will illustrate how an analog random signal generator can be generated from a thyratron noise source and a radioactive decay, and how a digital random signal generator can be generated utilizing a digital feedback shift register.

Heterodyne Technique

This method consists of utilizing a narrow-band noise signal that is heterodyned to a low frequency range.[18] For example, in Reference 18, the application of this technique for generating a 2-KHz noise signal is discussed. The output from a thyratron noise source is amplified by a 2-KHz selective amplifier. This noise signal is mixed with a 2-KHz sine wave. The statistical characteristics of the resulting low-frequency noise signal are greatly dependent on the noise properties of the thyratron noise source. In addition, its amplitude depends considerably on the amplification properties of the amplifiers, and stability is a great problem with this type of low-frequency random noise generator.

Noise Generation Utilizing a Radioactive Source[17]

The second low-frequency random noise generator utilizes a radioactive source in conjunction with a Geiger-Mueller (GM) type of tube. The output from the GM tube, which is triggered by the radioactive source, consists of a series of random pulses having a Poisson probability distribution given by:

$$p(x) = \frac{(\lambda t)^x e^{-\lambda t}}{x!}, \qquad (5.27)$$

where $p(x)$ is the probability of x pulses occurring in time T, and λ is the average occurrence rate of pulses per unit time.

Figure 5.20 illustrates the block diagram of the overall configuration. A radioactive source triggers the GM tube, which then triggers a bistable multivibrator. The output of this system is switched between voltage levels

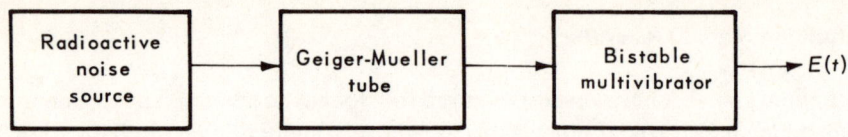

Figure 5.20. A Random Signal Generator Utilizing a Radioactive Source.[17]

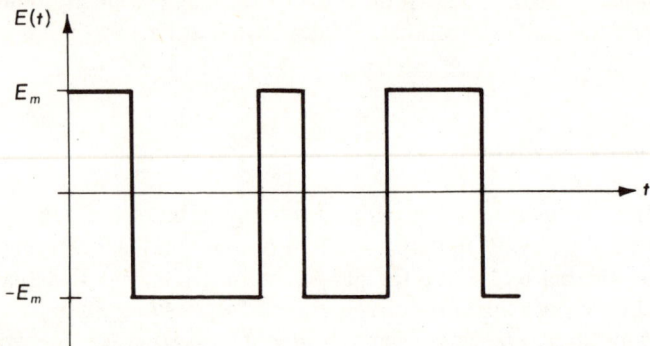

Figure 5.21. Output Waveform of a Radioactive Source

$\pm E_m$ at an average rate of K per second, as illustrated in Figure 5.21. Rice [19] refers to this random waveform as a "random telegraph signal." It can be shown that the power-density spectrum of the output $\Phi(\omega)$ is given by[19]

$$\Phi(\omega) = \frac{E_m^2(2K)}{\pi[(2K)^2 + \omega^2]}. \qquad (5.28)$$

Its statistical properties are illustrated in Figure 5.22.

By passing this signal through a low-pass filter, we obtain a probability distribution which approximates a Gaussian distribution. Figure 5.23 illustrates this process. This resulting low-frequency, random noise generator has fairly good stability, but its statistical properties are somewhat dependent on the characteristics of the noise generator.

Pseudo-Random Sequence Generator Utilizing a Feedback Shift Register[20,21]

The final generator considered in this section produces a random noise signal utilizing digital techniques. Instead of using a physical noise source such as a thyratron or a radioactive device, a digital technique utilizing a feedback shift register is employed.

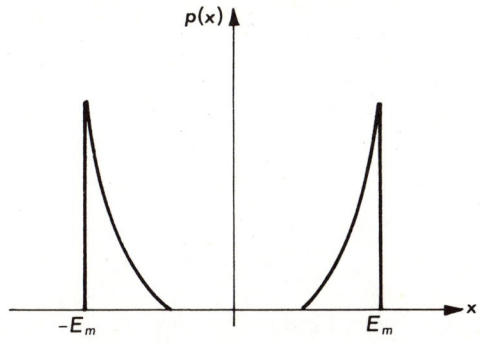

(a) Probability Distribution Function

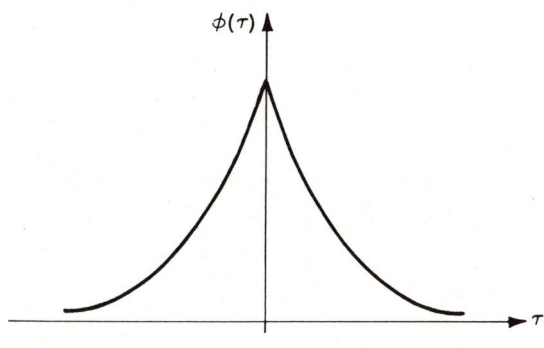

(b) Autocorrelation Function

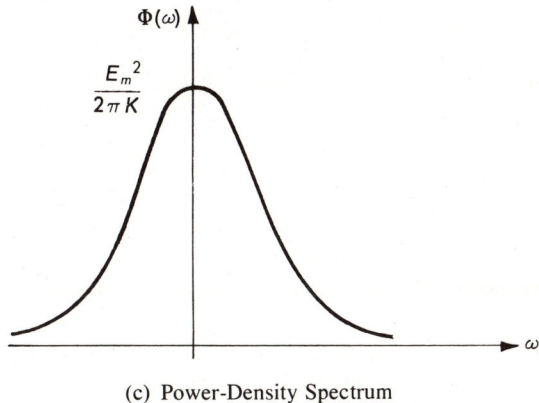

(c) Power-Density Spectrum

Figure 5.22. Statistical Characteristics of the Random Telegraph Signal

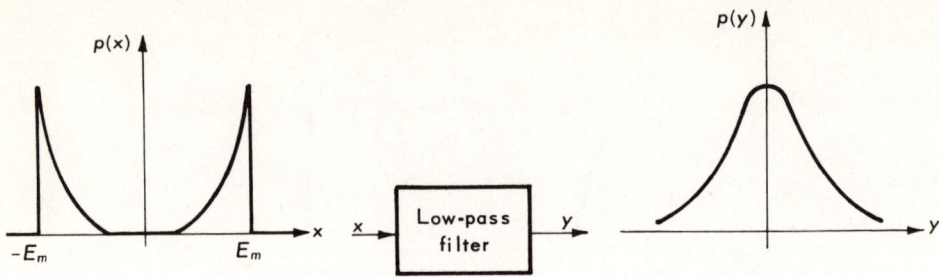

Figure 5.23. Conversion of the Random Telegraph Signal to a Low-Frequency Random Test Signal Having a Gaussian Distribution.

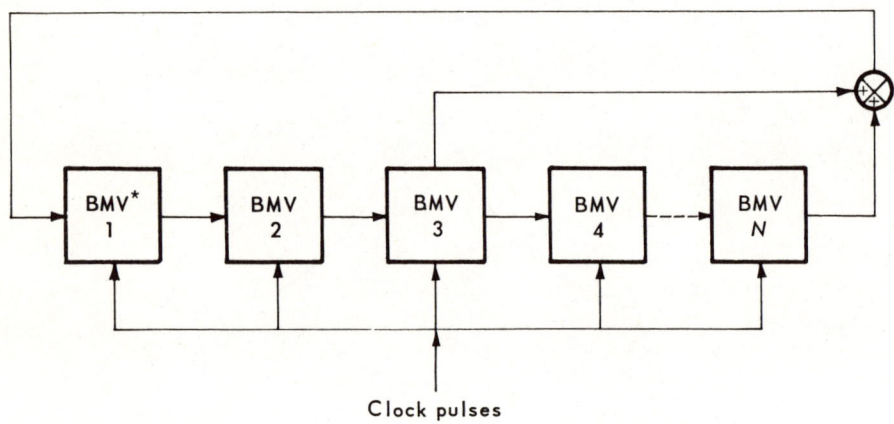

*Bistable multivibrator

Figure 5.24. A Feedback Shift Register

A feedback shift register, illustrated in Figure 5.24, is a binary circuit consisting of bistable multivibrators and "exclusive OR" gates.[20] The information contained in the chain is shifted one step forward upon the application of an input pulse.

Simultaneously, a bit obtained by a parity check on the information contained in several of the multivibrators, is shifted into the first bistable multivibrator. Therefore, a cyclic code circulates through the shift register. By properly designing the feedback shift register containing N bistable multivibrators, this code will contain all possible combinations of N bits

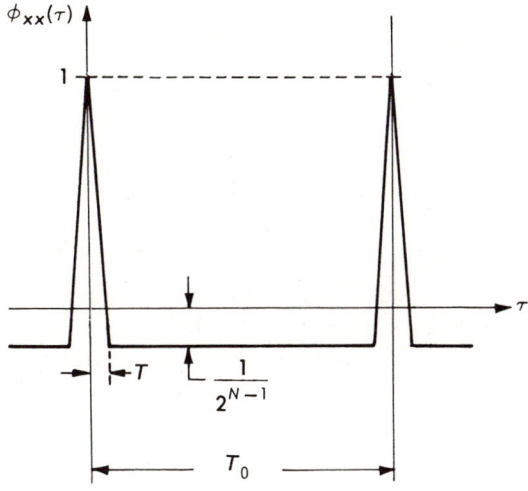

Figure 5.25. Autocorrelation Function for Shift Register Code

except when all bits are positive. The codes of these maximal length shift registers have a length of $2^N - 1$ bits.

When the code from the feedback shift register is multiplied by a time-shifted duplicate function of itself, the resulting code is the same code with a different time shift. This characteristic can be used to calculate its autocorrelation function. Since the number of negative bits exceeds the number of positive bits by one, the autocorrelation function is given by:

$$\Phi_{xx}(\tau) = \frac{1}{T_0}\int_0^{T_0} x(t)x(t+\tau)dt = -\frac{T}{T_0} = -\frac{1}{2^N - 1}. \quad (5.29)$$

The autocorrelation function for the cases when $\tau = 0, T_0, 2T_0, 3T_0$, etc., is unity. Figure 5.25 illustrates the resulting characteristics of the autocorrelation function.

The resulting autocorrelation function is identical with that of a real random binary sequence except that the series is periodic instead of random, and the autocorrelation function between the peaks is a small negative value instead of zero. However, both differences can be made as small as desirable by choosing N sufficiently large. Therefore, a maximal length shift-register code can be utilized to generate a pseudo-random binary sequence.

Similar to the method discussed previously, this digital sequence can be converted into a random noise test signal by passing it through a low-pass filter. The resulting generator has good, stable, and predictable statistical characteristics which are necessary for a random signal test generator.

References

1. Korn, G.A. and T.M. *Electronic Analog and Hybrid Computers*. New York: McGraw-Hill Book Company, Inc., 1964.
2. Ashley, J. Robert. *Introduction to Analog Computation*. New York: John Wiley and Sons, Inc., 1963.
3. Bartee, Thomas C. *Digital Computer Fundamentals*, 2nd ed. New York: McGraw-Hill Book Company, Inc., 1966.
4. Truxal, John G., ed. *Control Engineers Handbook*. New York: McGraw-Hill Book Company, Inc., 1958.
5. Elgerd, Olle I. *Control Systems Theory*. New York: McGraw-Hill Book Company, Inc., 1967.
6. *Basic Language Reference Manual*. General Electric Information Systems Division, June 1965, rev. January 1967.
7. McCracken, D.D. *A Guide to FORTRAN Programming*. New York: John Wiley and Sons, Inc., 1961.
8. Amberntson, D.S. "Hybrid Solution Methods of Split Boundary-Value Problems." M.S. Thesis, Department of Electrical Engineering, University of Florida.
9. Shinners, Stanley M. "Which Computer . . . Analog, Digital or Hybrid?" *Machine Design* (January 21, 1971) pp. 104-111.
10. Bekey, G.A. and Karpulus, W.J. *Hybrid Computation*. New York: John Wiley and Sons, Inc., 1968.
11. Deutsch, Ralph. *System Analysis Techniques*. Englewood Cliffs, N.J.: Prentice-Hall, Inc., 1969.
12. Machol, R.E., ed. *Systems Engineering Handbook*. New York: McGraw-Hill Book Company, 1965.
13. Chestnut, H. *Systems Engineering Tools*. New York: John Wiley and Sons, Inc., 1965.
14. Skolnick, Merril I. *Introduction to Radar Systems*. New York: McGraw-Hill Book Company, Inc., 1962.
15. Vander Velde, W.E. "Make Statistical Studies on Analog Simulation." *Control Engineering* (June 1960).
16. Uspensky, J.V. *Introduction to Mathematical Probability*. New York: McGraw-Hill Book Company, 1937.
17. Shankar, T.N. Shiva. "A Low-Frequency Random Noise Generator." *Journal of the Institute of Telecommunications Engineers*. 9 (1963).
18. Bell, D.A. and Rosie, A.M. "A Low-Frequency Noise Generator with Gaussian Distribution." *Electronic Technology*. 37 (1960) p. 241.

19. Rice, S.O. "A Mathematical Analysis of Random Noise." *Bell System Technical Journal*. 23 (1944) p. 282.
20. Huffman, D.A. The Synthesis of Linear Sequential Coding Networks, *Proceedings of 3rd London Symposium on Information Theory*, 1955.
21. Kramer, C. "A Low-Frequency Pseudo Random Noise Generator." *Electronic Engineering* (July 1965).

6 Man-Machine Control Systems: The Human Model

Introduction

Having presented the overall systems engineering problem and discussed various aspects of it in depth, in this chapter we concern ourselves with the role played by the human element in systems. The proper design of man-machine control systems requires as much understanding of the human element as that of the machine's characteristics. The modern systems engineer can obtain a very high degree of repeatability and accuracy for the machine's characteristics. However, the characteristics of the human element are not understood as well. With the complexity and importance of modern man-machine control systems, this difficult problem has received a lot of increasing attention.[1,2] This chapter presents a comprehensive analysis of human transfer functions that have been proposed, and provides guidelines for the proper design of man-machine systems.

Many modern systems depend on the performance of a human controller. Manual tracking systems, which date back to early World War II days, require operators to track targets based on certain visual information displayed. The interest in this aspect of engineering has been further stimulated by the importance of man-machine control systems in many aspects of modern command and control systems such as the air traffic control system discussed in detail in Chapter 1. In addition, the behavior of man has received further attention in the fields of manned space systems, aeronautics, machine tool control, biocybernetics, and other fields allied with medical electronics. This is a very important aspect of systems design which has received attention from engineers, scientists, psychologists, and medical doctors.

Man is an extremely complex system. The determination of a realistic mathematical model for the human controller, even for the performance of a specific task, is very difficult. However, in order to predict the stability and dynamic performance of systems which have human controllers, the systems engineer must know the mathematical input-output relationship of man.

The various attempts to describe the human controller's characteristics have all involved a number of simplifying assumptions. Most initial models were greatly in error due to the simplified models synthesized. As new analytic tools and knowledge became available to the control systems engineer, the models became more sophisticated and accurate. Although

133

the behavior of human controllers is better understood today than in the days of World War II, there is still much to be learned.

The unique characteristics of human controllers is presented first in this chapter, followed by the various models postulated by a great array of investigators. Since linear approximations are adequate in most systems applications, this chapter emphasizes linear and adaptive characteristics. Besides presenting the different concepts, comments are made on the usefulness of each model postulated. Emphasis is placed on the most recent literature contributions throughout the chapter. Guidelines are included for the application of the material presented to the proper design of man-machine control systems.

Unique Characteristics of Human Controllers

The human controller has several unique characteristics. His input-output relationships cannot be described as being purely linear, nonlinear, time-variable, random, or discrete. They are actually combinations of all these characteristics. In addition, man is a highly adaptive controller who learns from experience. Other factors which affect his performance are motivation, fatigue, and *a priori* information.

From a mathematical viewpoint, some of these unique characteristics can be described while others cannot. In the presentation to follow, attention is concentrated on inserting the following characteristics of the human controller's input-output behavior into a mathematical description:

Frequency Response

The human controller responds predominantly to low-frequency commands. He tends to attenuate high-frequency components where the attenuation increases with frequency.

Nonlinearities

Man's behavior indicates distinct nonlinear characteristics. However, some of his nonlinear characteristics may be linearized.

Time Variations

The human controller displays time-variable characteristics. This is due to his adaptive capability which permits him to change his characteristics with time, and his learning capability which changes his performance with time.

Discreteness

The human controller's response indicates that he behaves as a discrete, or sampled-data, system.

Delay

Man does not respond instantaneously and has a transportation lag associated with his response.

Randomness

The variations of a human controller's performance in successive trials indicates random characteristics. This variation can be reduced with adequate training, motivation, and simplicity of tasks.

Adaptiveness

Man is a highly adaptive control system, able to adjust his characteristics with a wide range of controlled element dynamics, and able to learn and predict future actions of an input function.

Although experimental data indicate that the human transfer function has the characteristics of nonlinearities, time-variableness, discreteness, randomness, and adaptiveness, most useful models that have been postulated are approximations which emphasize one or two particular characteristics. The results, therefore, between the synthesized model and experimental data are approximations. However, they do lend insight to the problem and have facilitated a better understanding of the problem today. These concepts have set the stage for the recent work that has included the effect of the human being in systems, and have resulted in better correlation between predicted and actual results.

Linear, Continuous Models of the Human Controller

The problem of adequately describing, mathematically, the behavior of the human controller dates back to the World War II period. Control systems engineers were concerned with the ability of a gunner to operate as part of a fast-acting fire-control system, and the ability of a radar operator to track a target from information presented on a visual radar display. These early studies simplified the problem by considering the operator as a quasi-linear system.

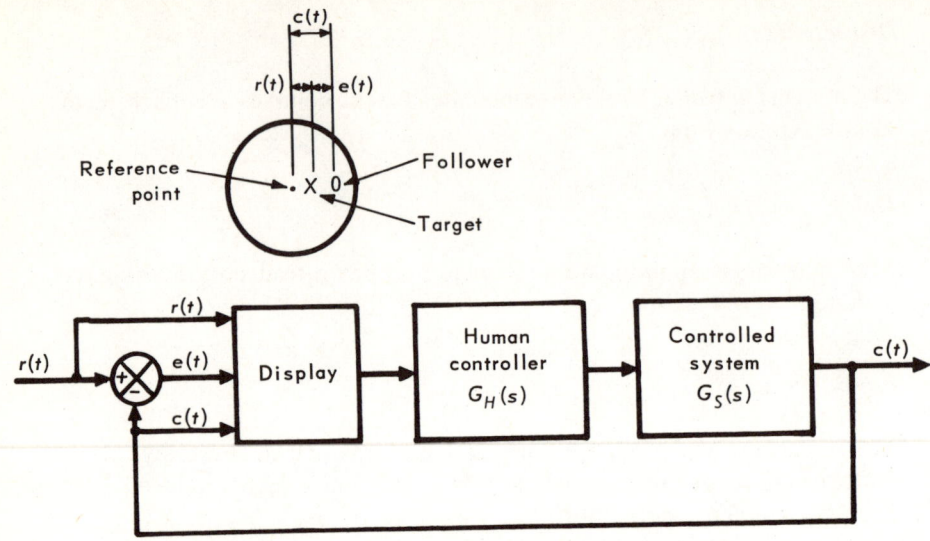

Figure 6.1. Pursuit Manual Tracking System

The manual tracking problem is used to introduce some of the early concepts regarding a human controller's characteristics. In general, tracking displays can be categorized as pursuit or compensatory. In pursuit manual tracking, as illustrated in Figure 6.1, the operator attempts to position a movable follower coincident with the independently moving target. In compensatory manual tracking, as illustrated in Figure 6.2, the operator attempts to maintain a movable follower coincident with a stationary reference which represents the target.

From the human controller's viewpoint, there are several important differences between pursuit and compensatory tracking. In pursuit tracking, the movable follower represents the system's output and is moved only in response to the target indicator's motion which represents the system's input. Comparison of input and output is done solely by the human operator. In compensatory tracking, the human operator sees only the difference between the system's output and input—namely, the error. He has no way of determining the target's actual position except by deducing it from his memory, the history of the error signal, and the history of his own previous movements.

Analysis of pursuit and compensatory tracking indicates that the greatest amount of useful information is made available in pursuit tracking. This advantage is very evident in the case of perfect tracking. For this case, a compensatory tracking display doesn't display any error signal or target motion information except data that the operator can infer from his own

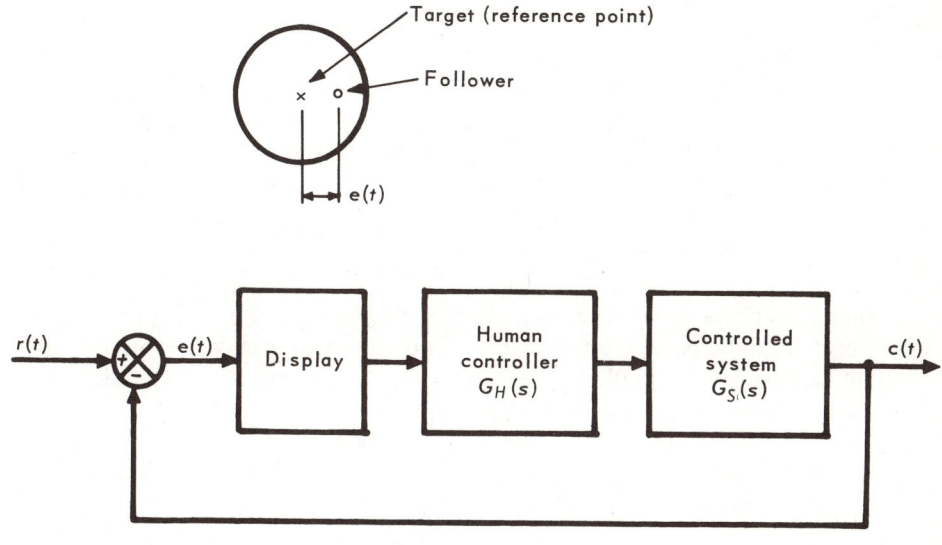

Figure 6.2. Compensatory Manual Tracking System

output. However, in the case of pursuit tracking, the operator has available information regarding the target's motion and his own output even though there is no error signal. In conclusion, therefore, it can be stated that a pursuit manual tracking system is preferable since more information is made available to the operator than in a compensatory manual tracking system.

During World War II, initial concepts regarding the human transfer function basically consisted of visualizing it as having linear, continuous characteristics. In 1943, R. S. Phillips[3] was concerned with the derivation of the human transfer function. This model was based on a step-function response of the form given by

$$G_H(s) = K\frac{(1 + T_A s)}{s}e^{-Ds}, \qquad (6.1)$$

where D represents the operator's transportation lag and T_A represents the operator's anticipation time constant. Phillips felt that the time constant T_A was very small and actually simplified the transfer function to

$$G_H(s) = \frac{K}{s}e^{-Ds}. \qquad (6.2)$$

In addition, he assumed that the operator's transportation lag, D, was 0.5 second.

Tustin[4] in 1947 theorized the human transfer function as a linear, continuous model having an additional disturbance of unknown origin which he called the "remnant." The remnant term, $N(s)$, basically consisted of the operator's response which was not linearly correlated with the input. A block diagram utilizing this concept is illustrated in Figure 4 of Mitchell's article.[1] The human transfer function as postulated by Tustin is given by

$$G_H(s) = \frac{K(1 + T_A s)}{s} e^{-Ds}, \qquad (6.3)$$

where D represents the operator's transportation lag and T_A represents the operator's anticipation time constant. This transfer function is identical to that synthesized by Phillips [see Equation (6.1)] except that Tustin did not assume the time constant to be very small.

Further considerations of the remnant term by recent investigators[1] have attributed it to one or more of four possible sources:

1. Noise at the operator's input
2. Noise at the operator's output
3. Unsteady behavior of the operator
4. Nonlinear operation and dither

In 1957, McRuer and Krendel[5] provided a very general and useful form of the human transfer function. It is given by:

$$G_H(s) = K \frac{(1 + T_A s) e^{-Ds}}{(1 + T_L s)(1 + T_N s)}, \qquad (6.4)$$

where D represents the operator's transportation lag, T_A represents the operator's anticipation time constant; T_L represents the operator's error smoothing lag time constant; and T_N represents the operator's short neuromuscular delay. The gain K, and the time constants T_A and T_L are usually considered to be variable according to the control task being performed. The transportation lag D and the time constant T_N are generally assumed to be fixed for each operator, but they are variable among operators within certain bounds. This transfer function has met with reasonable success in closed-loop tracking tasks utilizing compensatory displays. Representative values for the elements in Equation (6.4) are as follows:[1]

D = 0.2 second ±20%

T_A = 0 to 2.5 seconds (variable)

T_L = 0 to 20 seconds (variable)

T_N = 0.1 second ±20%

K = 1 to 100 (variable)

Analysis of these values indicates that the human transfer function is variable over a wide range and exhibits the characteristics of adaptivity. More will be said of this characteristic in subsequent sections of this chapter.

The linear, continuous models of the human transfer function give reasonably good correlation when an operator is well trained, motivated, performing simple tasks, and is tracking information of low-frequency content. However, this representation does not account for the experimental data which indicate that human behavior has characteristics of nonlinearity, time-variableness, discreteness, randomness, and adaptivity. Two of the major problems encountered are the nonlinearity and time-variableness of the human transfer function. These two factors actually account for a major portion of the "remnant" as referred to by Tustin. Actually, the size of the remnant is often very large and is usually the major portion of the actual response.

Linear, Discrete Models of the Human Controller

The continuous linear, nonlinear, and time-variable models presented so far in this chapter do not account for the discrete behavior of a human controller's actions. Only relatively recently has a serious attempt been made to synthesize discrete models for the human transfer function, although the presence of discontinuous characteristics in manual tracking has been known since World War II. Motivation for this research has been stimulated by the development of sampled-data theory and the availability of analog and digital computers for real-time simulation of models too complex for analysis.

The argument that the human transfer function should be represented by a discrete model has support from psychological grounds in addition to actual experimental data. Psychological refractory considerations indicate that the human controller cannot make two successive responses to discrete stimuli in an interval less than a refractory period of approximately 0.5 second.[6] This characteristic suggests a discrete model which can perform tasks only at certain periodic intervals.

One of the first discrete mathematical models was proposed by J. D. North.[7] Basically, he took Tustin's original linear, continuous model and replaced with finite differences the derivatives in the differential operation describing Tustin's model. The resulting difference equations also included the remnant component which he represented as pulses of white noise. Unfortunately, North focused his attention on the system's behavior at the sampling instants only while the human controller's output was actually a continuous function. He could have accomplished this by inserting a data extrapolator such as a zero-order hold in the model. North's observations

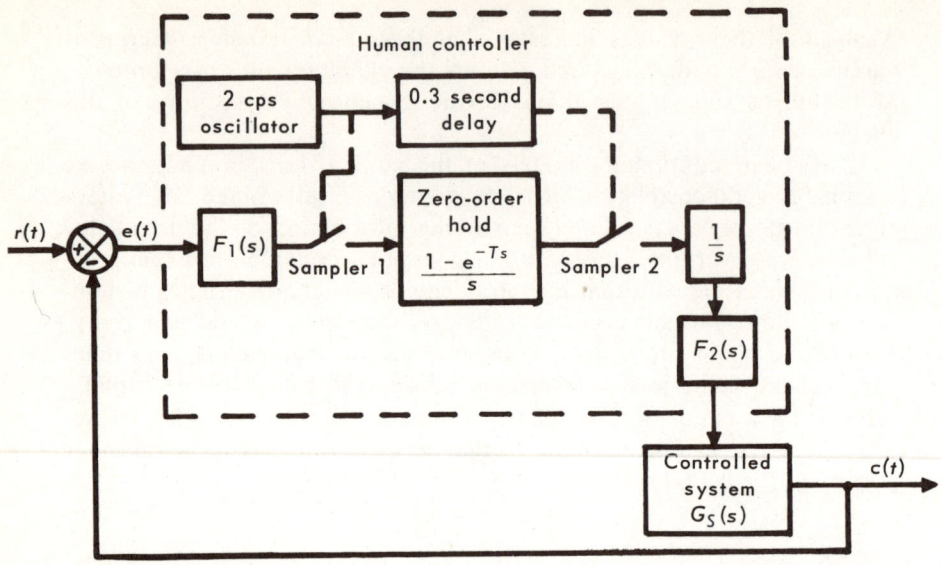

Figure 6.3. Ward's Discrete Model of the Human Controller as Redrawn by Bekey[7]

were valid only at integral values of sampling instants and therefore were not too revealing.

In 1958, Ward[8] proposed a discrete model of the human transfer function utilizing an analog computer. This model, as redrawn by Bekey,[7] is shown in Figure 6.3. The model's parameters were varied in an attempt to yield a good visual match for actual tracking data. The sampling rate chosen was 0.5 second. The transfer functions, $F_1(s)$ and $F_2(2)$, were determined empirically and are given by

$$F_1(s) = A + Bs + C \frac{s^2 + 2\zeta_1 \omega_0 s + \omega_0^2}{s^2 + 2(4\zeta_1)\omega_0 s + \omega_0^2} \cdot \frac{1}{1 + 0.5s} \quad (6.5)$$

$$F_2(s) = \frac{1}{1 + C \frac{s^2 + 2\zeta_1 \omega_0 s + \omega_0^2}{s^2 + 2(4\zeta_1)\omega_0 s + \omega_0^2} \cdot \frac{1}{1 + 0.5s}}. \quad (6.6)$$

The input forcing functions used to treat the model consisted of 8 sine waves ranging in frequency from 0.01 cps to 0.2 cps. Unfortunately, the model did not perform well in fitting either time or frequency domain data. Part of the problem can be traced to the integrator in the model that includes a sampler and zero-order hold followed by a delayed sampler, which accounts for reaction time, and the integrator. This integrator, however, is

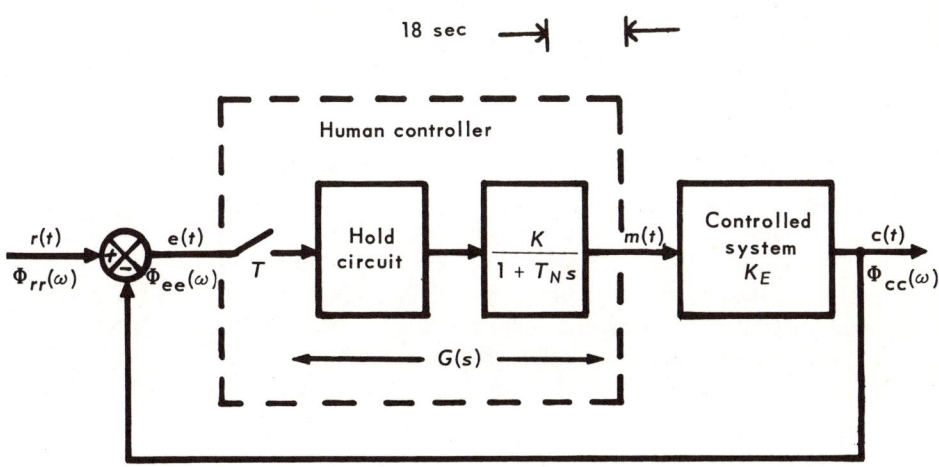

Figure 6.4. Bekey's Discrete Model of the Human Controller[9]

a nonresetting hold circuit and the reconstructed signal, R, has no resemblance to the continuous input.

Bekey's[9] model of the human transfer function in a compensatory tracking loop is shown in Figure 6.4. It includes a sample and hold circuit, and a neuromuscular delay time constant. An important contribution of Bekey was his recognition of the importance of the type of hold circuit used after the sampler. He found that a first-order hold circuit was more appropriate than a zero-order hold circuit using only the frequency domain characteristics as a criterion. Using a first-order hold, Bekey was able to match the frequency response of the model with that of the human operator who was tracking random inputs over a very wide range of frequencies.

In the block diagram shown in Figure 6.4, K_E represents the gain of the external controlled system; K is the gain of the human controller; and T_N is the neuromuscular time constant. $G(s)$ represents the combined transfer function of the human controller's continuous portion. The postulated model was tested with random signals and the results were also compared with that of simple linear, continuous models. The results of these comparisons indicated that the two models gave similar responses over low frequencies, but Bekey's model gave much better results at high frequencies of operation.

The Human Controller's Adaptive Capability

A large amount of evidence indicates that the human controller is a very complex and highly perfected adaptive feedback control system. Many researchers have performed a wide variety of tests to substantiate this conclusion. However, the actual topology of the adaptivity synthesized so far is only a very simple approximation. In many respects, they are similar to the simple adaptive systems that have been synthesized for autopilot applications.[12] They scarcely approach the actual complex adaptive characteristics of the human controller. Nevertheless, the basic rudimentary adaptive topology that control engineers are familiar with do lend themselves to understanding better the characteristics of the human controller.

The various classes of adaptation present in human controllers will be discussed first. Some of the experimental evidence that actually indicates adaptive characteristics in the human controller will then be indicated. Finally, several adaptive models that have been proposed are presented.

Stark and Young[6] classify four general types of adaptation in human operators as

1. Input Adaptation
2. Task Adaptation
3. Biological Adaptation
4. Learning

and they are defined as follows:

Input Adaptation—the method by which the operator adapts a control policy appropriate to the characteristics of the input's forcing function.

Task Adaptation—the method of adapting to changes in the gain or dynamics of the controlled system.

Biological Adaptation—the process of adapting based on sensory phenomena.

Learning—the process of developing skills based on past experiences. Figure 6.5 illustrates the basic characteristics of input adaptation, task adaptation, and biological adaptation. Observe that input adaptation basically involves the recognition of repeatable patterns of the input which result in changes of the control from compensatory to precognitive tracking. To illustrate learning, changes in the topology of the model are required rather than merely changing the control law.

Input Adaptation

Experimental evidence indicates that the operator can recognize periodic components of the input signal and use his *input-adaptive* characteristics to predict future characteristics of the input in order to improve his response. This form of adaptation has been studied by Tanner and Swets[10] from the viewpoint of statistical detection theory, by Elkind[11] from the viewpoint of

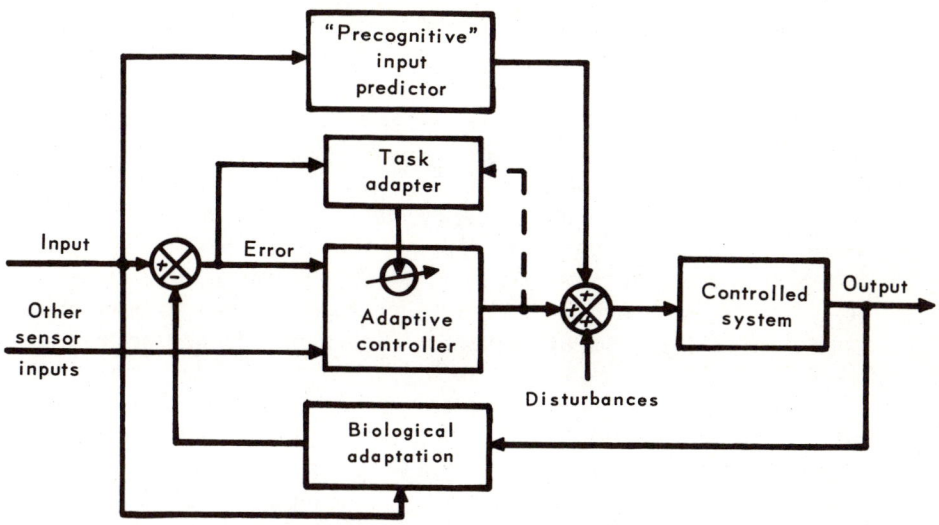

Figure 6.5. Various Types of Adaptive Characteristics Present in the Human Controller[6]

variable input bandwidth, and by Sheridan[12] from the viewpoint of tracking "clean" sinusoids. The results of their investigations are summarized in the following paragraph.

Most visual tracking data consist of signal and noise. Tanner and Swets[10] indicated that the human operator can detect the presence of a predictable signal in an input composed of signal and noise components. Elkind[11] investigated the adaptability of the human operator to changes of the input's bandwidth. He found that the human controller behaved differently in tracking low-frequency data than in tracking high-frequency data. Elkind found that the operator adapted himself to obtain a "tighter" tracking loop for low-frequency data. For higher frequencies, such control introduced excessive phase-lag and the operator had to adjust his transfer function by accentuating the high frequency response, reducing his gain at low frequencies, and changing his lag time constant. Sheridan[12] studied the problem of an operator in a compensatory tracking situation following a "clean" sinusoidal signal. Experiments indicated that the human operator can track the sinusoidal signal almost perfectly and he changes to a precognitive tracking mode after properly identifying the signal. These examples clearly indicate that the human controller adapts a control policy appropriate to the characteristics of the particular input forcing function and he exhibits "input adaptation."

Task Adaptation

A large amount of experimental evidence also indicates that the human controller adjusts himself to changes in the dynamics or gain of the controlled system and exhibits the characteristic of *task adaptation*. Sadoff,[13] McRuer and Krendel,[14] and Elkind, Kelly, and Payne[15] have studied the capability of the human operator for adapting to changes in the controlled system's dynamics. Young, Green, Elkind, and Kelly[16] have also studied the capability of the human operator for adapting to changes in the gain of the system. As indicated in the following paragraphs, experimental evidence clearly supports the belief that the human controller has the characteristic of "task adaptation."

Human pilot adaptation to sudden changes in the dynamics of the controlled system were studied by Sadoff.[13] He experimented with tests to determine a pilot's capability to control the pitch axis of a moving flight simulator during a failure of the auxiliary pitch damping system. The effect of this failure on the system was to decrease the effective damping constant of the second-order system from 0.3 to 0.04, while leaving the undamped natural frequency constant. The results of this experiment are indicated in Figure 6.6. The recordings indicate that, initially, the pilot was attempting to test the reaction of the pitch loop, and this was followed by about 25 seconds of tracking of low-frequency random input. When the damper failed, the loop went into an oscillation of approximately 1 cycle per second for about 18 seconds. The recordings also indicate that the pilot was subjected to peak angular accelerations in pitch as large as 60 degrees per second squared. After the errors were finally reduced, steady-state tracking was poorer than when the pitch damper was in the loop. However, the pilot did demonstrate his capability to adapt to changes in the dynamics of the controlled system.

In 1962, McRuer and Krendel[14] hypothesized that the human operator is capable of changing his dynamics to compensate for changes in the dynamics of the controlled system. Considering the linear, continuous general form of the human transfer function to be [see Equation (6.7)][5]

$$G_H(s) = \frac{K(1 + T_A s)e^{-Ds}}{(1 + T_L s)(1 + T_N s)} \qquad (6.7)$$

McRuer and Krendel indicated that K, T_A, and T_L were functions that the operator varied adaptively. This characteristic was implied previously in this chapter. Lags in the system may be compensated for by increasing the time constant, T_A. Similarly, by increasing T_L, the operator can filter out high-frequency noise. The value of K can be adjusted by the operator to vary the bandwidth and/or the degree of stability. The general form of the adaptive model proposed by McRuer and Krendel is illustrated in Figure

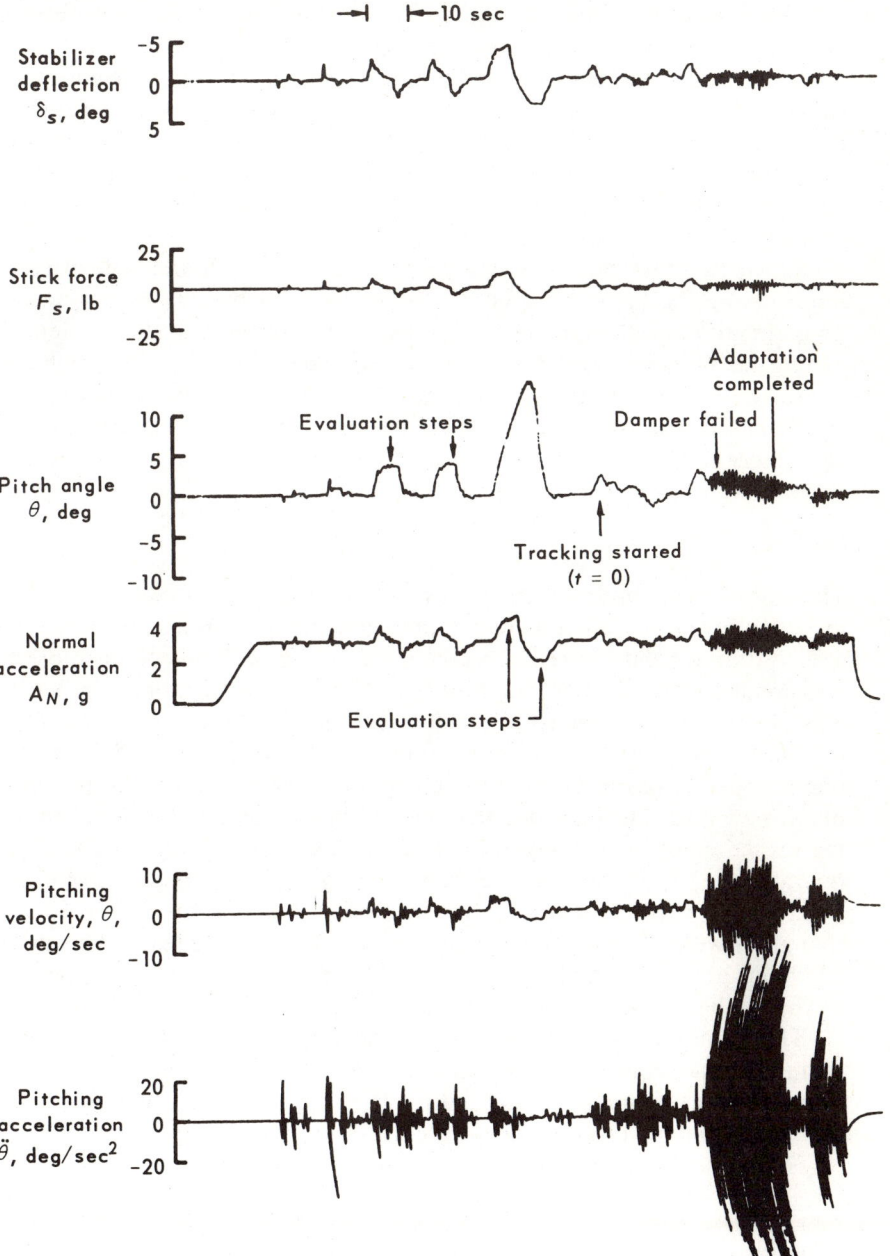

Figure 6.6. History of a Pitch Damper Failure[13]

6.7. They have separated the overall system into components covering visual sensor delay, data processing, compensation, synaptic and conduction delays, and the neuromuscular system which accounts for the time constant T_N. The entire neuromuscular system can be approximated at low frequencies by the transfer function $1/(1 + T_N s)$. The limits of the various transportation lags are indicated in Figure 6.7. Best results were found when the gain K was adjusted to yield an overall system phase margin between 40 to 80 degrees.

McRuer and Krendel's model of Equation (6.7) has found wide use for representing the human model in modern man-machine system applications. In addition, the limits given by Figure 6.7 are very useful for determining the limits of the human ability to respond in manual system applications. The reader should first try to use McRuer and Krendel's relatively simple model in most systems applications before more sophisticated models are tried.

Learning

The learning capability of the human operator has been studied by many investigators. It is well known that an operator improves his performance as he is trained for the job he is to perform. A very good study of this subject was conducted by Krendel and McRuer.[17] The explanation of this process was postulated on a model for skill development based on "successive organizations of perception." Krendel and McRuer assumed that the operator passes through different topological phases in the learning process of any new task. They claimed that the first phase consisted of concentrating on the error, as displayed by a compensatory tracking system. The operator then becomes capable of predicting the input as he recognizes certain aspects of his response. He uses this stored information to act as though he were in a pursuit tracking system. After having learned his task and being able to recognize certain inputs, the human controller responds as though he is tracking precognitively.

This summary of experimental evidence clearly supports the premise that the human controller has adaptive characteristics which are sensitive to changes in the input, tasks, and his learning capability, in addition to biological factors. The topology for representing these adaptive characteristics are considered in the following section.

Adaptive Representation of the Human Controller

The concept of adaptive control has been applied to autopilots.[12] Its application to the representation of the very complex characteristics of the human controller is not an easy problem. Most of the work done in this

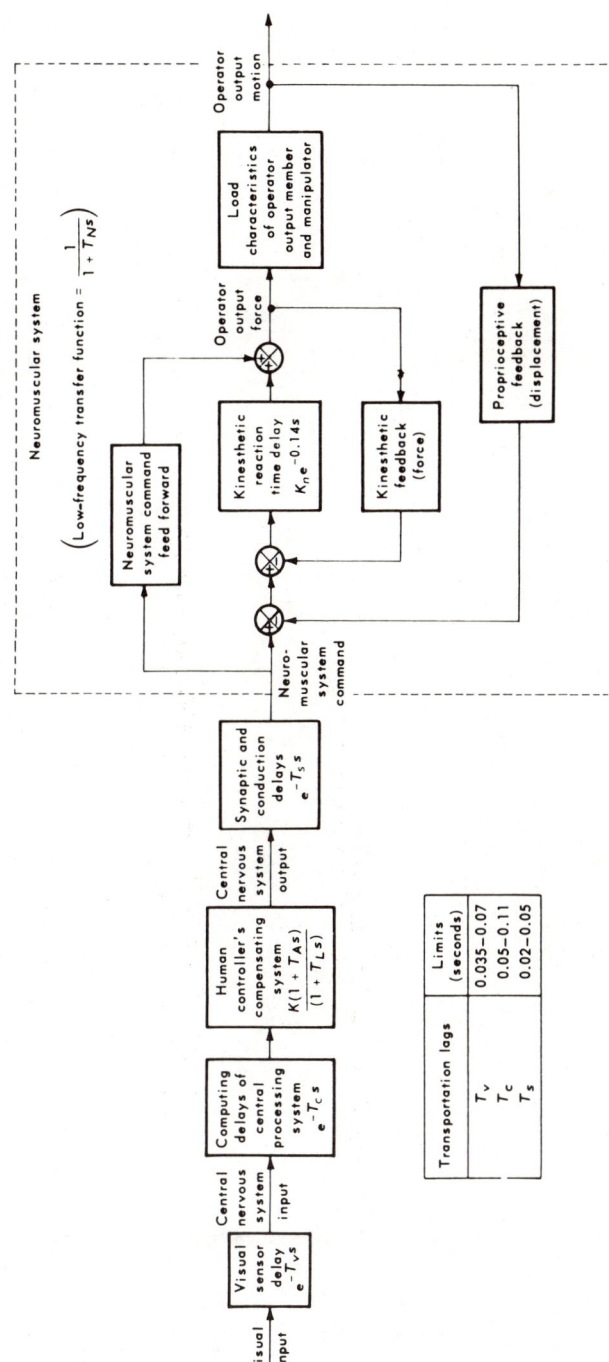

Figure 6.7. Adaptive Model of Human Operator Proposed by McRuer and Krendel[14]

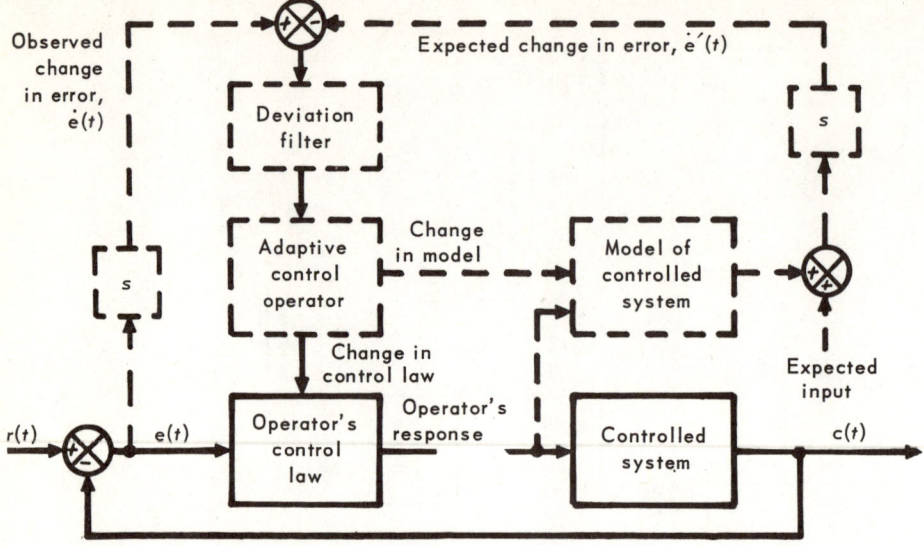

Figure 6.8. "Model Reference" Adaptive Control System for the Human Controller[6]

particular area recognizes the difficult problems involved and offers, instead, a set of guidelines, hypotheses, and constraints to be considered in the actual task of formulating a representative model. Two of the more interesting and promising ideas are presented in this section.

Young and Stark[6] propose two models for describing the adaptive processes concerned with compensatory tracking. They are referred to as the "model reference" and the "error pattern recognition model" which are illustrated in Figures 6.8 and 6.9, respectively. The dotted lines in these figures correspond to hypothetical paths of the adaptive process.

The basic element for the model reference concept is the device labeled "model of controlled system." It represents the human controller's concept of what the output should be, based on the response of the system to his input. The difference between the observed change in error, $\dot{e}(t)$, and the expected change in error, $\dot{e}'(t)$, is fed to the "deviation filter" which sets a threshold to reject insignificant deviations. The "adaptive control operator" varies the "operator's control law" in accordance with signals received from the deviation filter. The resultant process causes the operator to adapt his control laws to compensate for changes in the controlled system. In addition, the adaptive control operator also updates the operator's hypothetical model of the controlled system.

The error pattern recognition concept is based on the assumption that

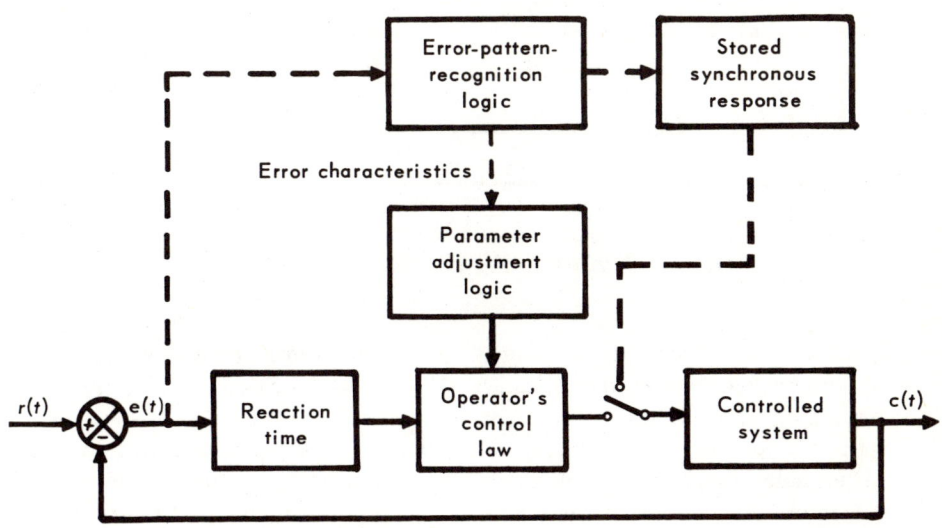

Figure 6.9. "Error Pattern Recognition Model" for the Human Controller[6]

the only information used by the operator in his adaptive processes is the error displayed. In this adaptive concept, the operator searches for error patterns to detect changes in the controlled system as reflected in changes of the closed-loop response. Using this procedure, the operator forms and stores a programmed response which is used in a precognitive tracking process. From Figure 6.9, it can be seen that this type of tracking bypasses the operator's inherent reaction-time element.

Techniques for Aiding the Human Controller

Although the human controller is a very versatile, complex, and adaptive mechanism, he does have limits to the degree with which he can control the operation of closed-loop systems. For example, the human operator has limitations in his ability to generate sufficient lead for overcoming the lags present in controlled systems having more than two integrations. To help the human controller perform his tasks, "aided tracking" or "quickening" techniques have been devised. These methods are presented in this section, together with comments on their usefulness.

Aided tracking techniques can be implemented in two general ways. They can be incorporated into a system by adding information to the

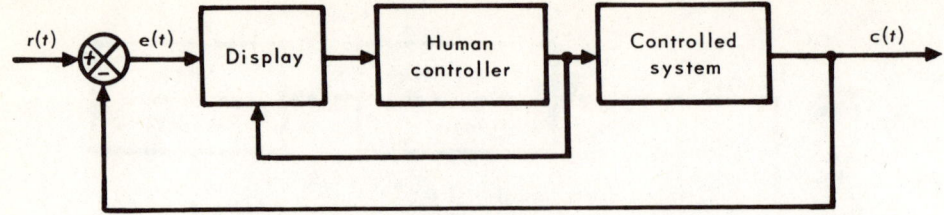

Figure 6.10. Aided Tracking Obtained by Adding the Human Controller's Output Directly to the Display

display or by synthesizing control signals that are derivatives of the operator's basic position commands. When aided tracking is incorporated by adding information to the display, its primary effect is to relieve the human controller of supplying smoothing and prediction information. When aided tracking is implemented by synthesizing system commands, which are derivatives of the operator's position commands, the primary effect is to assist the operator in supplying lead in order to compensate for the lags of the controlled system.

In general, there are three types of methods for adding information to the display. These are as follows:

1. Addition of derivatives of the system output to the display.
2. Direct addition of the human controller's output to the display (see Figure 6.10).
3. Addition of the states of the controlled system's dynamics to the display (see Figure 6.11).

There are two basic techniques for adding these quantities to the display. One simple way is to display the quantities separately. Another method is to add them directly to the system output or error before the signal reaches the display. In this manner, the motions of the display signals are appropriately modified. In either case, the net result is to relieve the human controller of the need to supply smoothing and prediction in his tracking task.

The basic compensatory manual tracking position loop illustrated in Figure 6.2 can be modified to a rate-aided tracking and rate-acceleration-aided tracking configuration by synthesizing system commands which are derivatives of the operator's position commands. Figure 6.12 illustrates the general block diagram of the rate-aided tracking configuration,[18] and Figure 6.13 illustrates the general block diagram of the rate-acceleration-aided tracking configuration. These systems use the discrete model of the human controller as proposed by Adelman and Shinners.[18] The differentiators shown may be electronic or electromechanical devices. In rate-aided tracking, the human controller varies the output position directly and, indirectly,

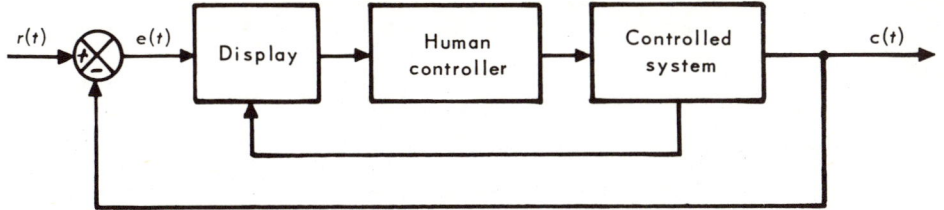

Figure 6.11. Aided Tracking Obtained by Adding the States of the Controlled System's Dynamics to the Display

also changes the output's speed by means of the synthetically derived rate signal. In rate-acceleration-aided tracking, the situation is very similar to that of rate-aided tracking except that a synthetic acceleration signal is also derived. Therefore, the human controller actually is varying position, velocity, and acceleration when adjusting his manual position control. The primary effect of these aided tracking configurations is to assist the operator in supplying lead to compensate for the lags in the controlled system.

A convenient parameter for expressing the relative degree of the synthetically derived signal used in aided tracking configurations is the aided-tracking ratio (ATR). For the rate-aided tracking case, the ATR expresses the ratio between output displacement and output velocity. For example, an ATR of 1:3 means that for every 1 degree displacement of the output position, an output velocity of 3 degrees per second is added simultaneously to the output. This is also referred to as an aided-tracking time constant of 0.33 second. For the rate-acceleration-aided tracking case, the ATR is expressed as the ratio of output displacement, to output velocity, to output acceleration. For example, an ATR of 1:3:6 means that for every 1 degree displacement of the output position, an output velocity of 3 degrees per second and an output acceleration of 6 degrees per second squared are added simultaneously to the output.

Various values of ATRs have been reported by several investigators. For the rate-aided tracking case, most of the values lie between 3:10 to 7:10.[19] However, values above 7:10 have been suggested for very slow moving targets[20] and values below 3:10 have been suggested for very fast moving targets.[21] For the rate-acceleration-aided tracking case, Searle[22] concluded that the best ATR was 1:4:8.

During the introductory discussions of pursuit and compensatory manual tracking systems on page 137, it was stated that the greatest amount of useful information was made available to the operator in pursuit tracking and therefore was preferable. However, when aided tracking is considered, compensatory tracking systems appear to be preferable. Anderson et al.[23]

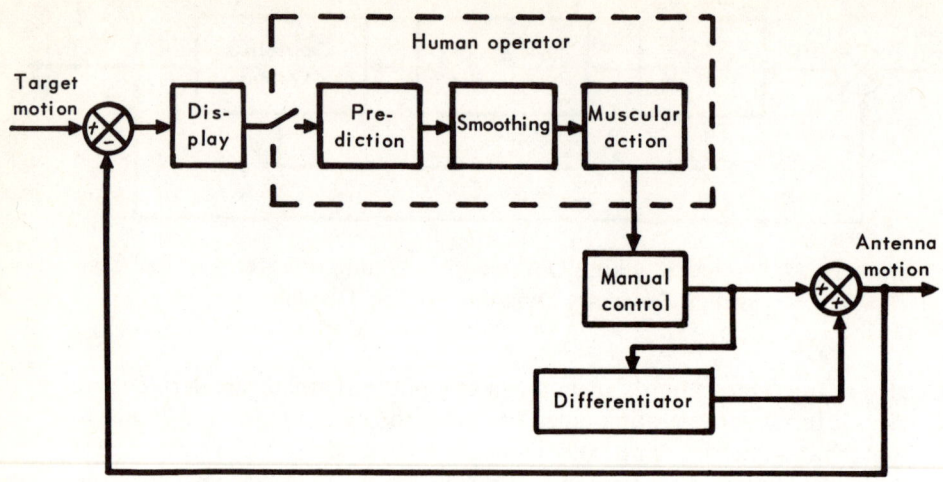

Figure 6.12. A Rate-Aided Tracking System[18]

found that compensatory manual aided tracking systems yielded slightly better performance than pursuit manual aided tracking systems. Other investigators appear to agree on this point within varying degrees.[24]

Although the techniques described in this section have proven somewhat useful for aiding the operator, they are very crude and probably very far from optimum. The main reasons for this are that the exact topology of the human control system and the exact transfer function of the human controller are not known. Until they are completely known, control engineers cannot completely and optimally devise systems to aid the human operators. Very basically, this problem is very closely related to the identification problem of control systems.

Conclusions

At this stage of the development of this chapter, the reader has been presented with a great variety of human transfer function concepts. Linear continuous and discrete, and adaptive models, postulated by several investigators, have been presented. In addition, characteristics unique to human controllers have been presented. All of the models presented account for certain unique characteristics of the human controller, but none accounts for all of them. The question remains: What is the correct model of the human controller to be used by the systems engineer?

The best recommendation I can suggest is that the engineer consider

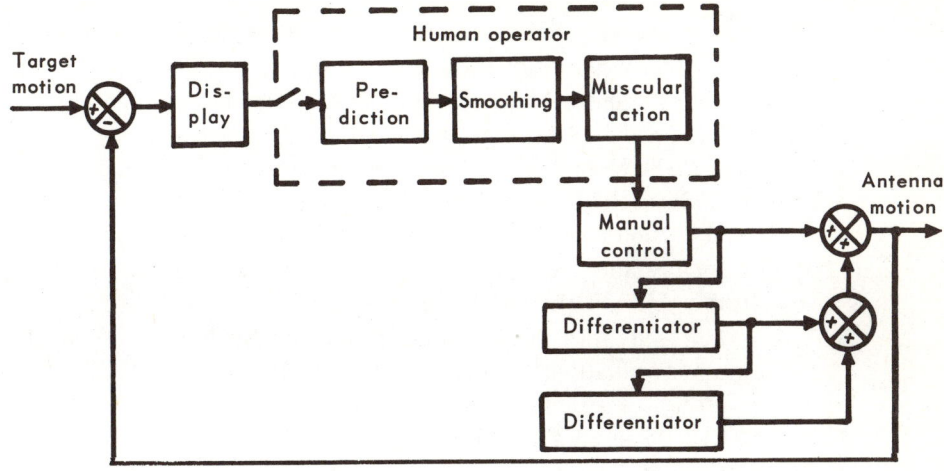

Figure 6.13. A Possible Rate-Acceleration-Aided Tracking Configuration[18]

each application separately and see the simplest model which will give good engineering results. For example, if the problem concerns the manual tracking of a relatively slowly moving target and the frequency spectrum of the system has a relatively low bandwidth, use the simple approximation given by McRuer and Krendel[5] [see Equation (6.4)]. On the other hand, if the problem involves adaptation on the part of the operator, the model postulated by Bekey[9] (see Figure 6.4) may be considered. Between these two extremes are all the other models postulated. The only reasonable advice is to consider each problem separately and determine which model fits the problem best.

Before leaving the subject of the human controller, here's a word about other factors which should be considered in a realistic situation. Any of the models proposed for the human operator are worthless if the operator is suffering from fatigue and/or is not motivated. Best results are obtained with operators who are motivated in doing their tasks and are not fatigued. In addition, the operator's performance can be further improved if certain *a priori* information regarding what he should expect during his mission is given to him before the task begins.

References

1. Mitchell, Meredith B. "Systems Analysis—The Human Element." *Electro-Technology* (April 1966).

2. Halsted, "Improving the Information Flow Rate Between Man and Machine." *Electronic Industries*, (April 1966).
3. Phillips, R.V. "Servo Mechanisms." M.I.T. Radiation Laboratory, Report 372 (1948).
4. Tustin, A. "The Nature of the Operator's Response in Manual Control and Its Implications for Controller Design," *Journal of the Institution of Electrical Engineers* 94 (1947) pp. 190-202.
5. McRuer, D.T. and Krendel, E.S. "Dynamic Response of Human Operators." USAF, WADC TR 56-524, 1957.
6. Young, L.R. and Stark, L. "Biological Controls Systems—A Critical Review and Evaluation." *NASA Contractor Report*, NASA CR-190, March 1965.
7. North, J.D. "The Human Transfer Function in Servo Systems." In *Automatic and Manual Control*, edited by A. Tustin, pp. 473-502. London: Butterworth, 1958.
8. Ward, J.R. "The Dynamics of a Human Operator in a Control System: A Study Based on the Hypothesis of Intermittency." dissertation, Department of Aeronautical Engineering, University of Sydney, Australia, 1958.
9. Bekey, G.A. "Sampled Data Models of the Human Operator in a Control System." dissertation, Department of Electrical Engineering, UCLA, 1962.
10. Tanner, W.P. and Swets, J.A. "A Decision Making Theory of Visual Detection." *Psychological Review* 61 (1954) pp. 401-409.
11. Elkind, J.I. "Characteristics of Simple Manual Control Systems." TR 111, Lexington, Mass.: M.I.T. Lincoln Laboratories, 1956.
12. Sheridan, T.B. "The Human Operator in Control Instrumentation." *Progress in Control Engineering* 1. London: Hayward & Co., 1962.
13. Sadoff, M. "A Study of a Pilot's Ability to Control During Simulated Stability Augmentation System Failures." NASA TN D-1552, 1962.
14. McRuer, D.T. and Krendel, E.S. "The Man-Machine Concept." *Proceedings of the IRE* 50 (1962) pp. 1117-1123.
15. Elkind, J.I., Kelly, J.A., and Payne, R.A. "Adaptive Characteristics of the Human Operator in Systems Having Complex Dynamics." *Proceedings of the Fifth National Symposium on Human Factors in Electronics*. (May 1964) pp. 143-159.
16. Young, L.R., Green, D.M., Elkins, J.I., and Kelly, J.A. "The Adaptive Dynamics Response Characteristics of the Human Operator in Simple Manual Control." NASA TN D-2255, 1964.

17. Krendel, E.S. and McRuer, D.T. "Servo Approach to Skill Development," *Journal of the Franklin Institute* 269 (1960).
18. Adelman, S. and Shinners, S.M. "Radar Tracking Utilizing Operational Dynamics Regeneration." 1962 IRE International Convention, New York, 1962.
19. Krendel, E.S. "Design of Tracking Devices with Regard to Human Requirements." Department of Defense, Research and Development Board, ATI-193086, 1953.
20. Sobczyk, A. "Aided Tracking Supplement." M.I.T. Radiation Laboratory Report 452 (1943).
21. Gottsdanker, P.M. and Biel, W.C. "A Study of Tracking on Directors M-5-A2 and M-5-A2E1." *OSRD* 5929, 1945.
22. Searle, L.V. "Psychological Studies of Tracking Behavior, Part 4. The Intermittency Hypothesis as a Basis for Predicting Optimum Aided-Tracking Time Constants." NRL Report 3372, ATI-134-251, 1951.
23. Anderson, I.H., Baldwin, A.L., et al. "Comparison of Range Tracking Methods: Tracking to a Fixed Hairline vs. Tracking to a Rotating Hairline." *OSRD* 3358, ATI-14 469, 1944.
24. Chernikoff, R., Birmingham, H.P., and Taylor, F.V. "A Comparison of Pursuit and Compensatory Tracking Under Conditions of Aiding and No Aiding." *Journal of Experimental Psychology* 49 (1955) pp. 55-59.
25. Holden, Frank M. and Shinners, Stanley M. "Identification of Human Operator Performance Models Utilizing Time Series Analysis." *Proceedings of the Ninth Annual Conference on Manual Control*. Cambridge, Mass.: Massachusetts Institute of Technology (May 24, 1973).
26. Shinners, Stanley M. and Berger, Charles R. "Investigation and Identification of Human Operator Performance Models Using the Theory of Time-Series Analysis." AMRL-TR-72-88, February 1973.
27. Kuo, Benjamin C. *Automatic Control Systems*. Englewood Cliffs, N.J.: Prentice-Hall, Inc., 1972.
28. Shinners, Stanley M. "Modeling of Human Operator Performance Utilizing Time Series Analysis." *IEEE Transactions on Systems, Man and Cybernetics*. SMC-4 (September 1974) pp. 446-458.

7

Testing Techniques

Introduction

The testing of a complex system is a very important part of the overall systems engineering problem. It is an important process in the overall engineering of a modern system that is concerned with verifying previous assumptions, diagnostic analysis, customer acceptance, and confirming manufacturer's data. In this chapter, techniques are presented for correlating the actual characteristics of a physical system with anticipated theoretical values.

System test is performed to provide answers to one or more of the following questions:

1. Do the test inputs represent the intended conditions of usage with adequate realism?
2. When the specified inputs are applied, will the system produce the required outputs?
3. How does system performance differ from the design goal?
4. Are the system and its normal complement of personnel reliable enough to ensure that adequate results will be achieved at times of critical need?
5. Can personnel of the anticipated levels of skill operate and service this system adequately?
6. Does the system achieve its anticipated reliability?
7. Can the system and its personnel operate over the intended life of the program within the limits of anticipated costs?

The extent to which each of these questions enters into system testing depends upon the history and plans of each system. In practice, it is common to plan systems tests in graduated and distinctly different phases in order to answer these questions. Generally, the earlier tests are geared to answer specific questions and later phases are designed to answer the more general questions.

The term "test" is all-inclusive. Literally speaking, testing includes the phases of system development, system performance evaluation, environmental test, and life test. This chapter is concerned primarily with testing as it relates to systems development and performance evaluation. Environmental and life testing are very specialized fields. They are not unique to any system and are reserved for books specifically devoted to the subject.[1] However, it is assumed in this chapter that the tests are performed under realistic conditions which take into account such variations as line voltage and frequency.

When an element, or an entire system, is subjected to test, the systems engineer must recognize the compatibility of the theoretical characteristics of the unit or system under test and the capabilities of the testing apparatus. It is very important that inherent characteristics of nonlinearity, time variability, discreteness, and noise all be recognized and accounted for in the unit or system. In addition, the capabilities and limitations of the test apparatus must be determined. It is essential that the test engineer recognize the importance of these two factors and make sure that the unit under test and the test apparatus are compatible.

All of the basic linear, nonlinear, statistical, sampled-data, and adaptive characteristics of the unit or system under test should be thoroughly understood by the test engineer in order to set up the test properly and interpret the results intelligently. In practice, the task can sometimes be simplified by focusing attention on only one aspect of the problem, such as linear gain. However, the problem can never be simplified to the point where the systems test engineer can forget everything else and focus his attention on only one aspect of the problem. The high degree of accuracy required of modern complex systems makes it mandatory that attention be given to all of the dynamic, random, and systematic errors present in the overall system.

The Basic Foundations of System Test[1,2]

The basic foundations of system test include a plan, schedule, organization, and facilities for accomplishing this task. System test must be incorporated during the early stages of the design and development effort of the project. This effort includes planning, scheduling, and organizing the overall system test. A well-planned and organized test program has a high probability of succeeding. However, a program conceived only as an afterthought has a very small chance of success.

The *overall system test plan* should consider the objectives of test, schedule requirements, organization requirements, operating philosophy, and liaison requirements with the customer. The overall system test plan should then be subdivided into *detailed subsystem test plans*. These detailed plans should contain the objectives of the test, test methods, successful criteria, test responsibilities, and schedule information.

Scheduling system test early in the program is extremely important. This should include component, subsystem, and formal system test. This schedule should also consider long lead time test instrumentation and test facilities. PERT, which was discussed in Chapter 4, is a very useful tool to be used by management to update and formally review test programs.

Organization of the system test team is a very important part of the overall test program. In most large organizations, the test of a large-scale system requires the support of several organizations. Typically, the engineering, test, reliability, quality control, and inspection departments are involved. It is important that a system test organization be established to indicate the areas of responsibility and liaison requirements for each department.

Finally, adequate *test facilities* are required for a successful program. The equipment required depends to a large extent on the particular system being tested. Modern test facilities are expensive and require long lead times to develop and construct. These factors must be considered very carefully in developing the overall system test scheme.

Computer-Controlled Testing[3]

The innovation of modern, large, flexible digital computers has permitted fast and accurate evaluation of digital and analog equipment. Automatic testing techniques utilizing digital computers require a thorough analysis of the system's performance characteristics. To accomplish this, data must be collected, stored, and then operated upon by a series of calculations to obtain a measure of performance.

A general-purpose digital computer can be utilized for diagnostic testing at the assembly and subassembly level. Basic subroutines can be stored in the memory and may be modified for special testing requirements. The digital computer allows the use of sophisticated mathematical analysis in order to isolate failures, and also allows for future development of new testing techniques.

The major problem facing the systems test engineer when utilizing a digital computer is the requirement for storing large amounts of data and instructions. A typical test program may consist of 50 tests with an average of 1,000 bits for each test. This results in a storage requirement of 50,000 bits. About 30 percent of these bits are typically used for communication in order to identify tests and the unit under test, give special instructions, and command the necessary repairs. There is a definite requirement in this area for more efficient computer communication.

Application to the Testing of Digital Systems

A digital computer is very applicable for testing digital-type systems. For example, a digital word containing several bits can be inserted into the computer memory from the digital system being tested. The digital computer can then perform a direct comparison of this word with the desired result. This type of test results in a computer answer in the form of greater-than, less-than, or equal to.

In the case of straight transfer-type digital systems, a shifting routine can be utilized to locate all false channels. The program can shift the data into an accumulator and analyze it by means of the logic decision instructions. Since the computer knows the correct data pattern, transfer can be made on positive or negative, depending on the desired result. When a fault is detected, the shift number can be stored in a predestined position in the memory. Therefore, the entire message can be analyzed and the results can be printed out listing all malfunctioning channels.

The digital computer also can be programmed to take a numerical difference of the digital system under test. Knowing what result should be desired for each specific problem, the actual test result is subtracted from the desired result to obtain a numerical difference. If the answer is not zero, the computer can then insert a series of problems into the device in order to isolate the malfunction.

The computer does not have to compare each message as it is generated. Instead, it may store the various responses in its memory at a speed limited only by its own memory cycle time. These results can then be analyzed after completing the high-speed data collection.

Application to the Testing of Analog Systems

By utilizing its computational capability, the digital computer also can be applied to the problem of testing analog systems. However, the computer cannot locate faults as readily as in digital systems, since many important performance characteristics which indicate malfunctions in analog systems, can be obtained only indirectly.

Digital computers can be programmed to perform many analytic determinations on analog systems which ordinarily would require much manual labor. For example, in the control system area, the root locus can be readily determined utilizing a digital computer. This is discussed in greater detail in Reference 6.

Computer Techniques for Analyzing Test Data[4]

The digital computer is a very powerful tool for evaluating test data. It can calculate standard deterministic and statistical characteristics, in addition to determining the characteristics of an arbitrary wave. Mathematical functions which closely "fit" the test data can be constructed very readily.

This aspect of digital computer utilization during system test is considered in this section, from the viewpoints of curve fitting and harmonic analysis.

Curve Fitting

A mathematical formula can be fitted to a curve obtained from measured data in several manners, utilizing a digital computer. However, it is usually accomplished in digital computers by means of the *least-squares method*. In general, a polynomial of the form

$$y(x) = A + Bx + Cx^2 + \ldots + Px^n \tag{7.1}$$

can be fitted to any curve provided that a sufficient number of terms are utilized.

A measure of the quality of the fit of the polynomial to the curve is determined by means of *residuals*. They are defined as the differences between corresponding computed polynomial values and the measured data values. For example, the residuals of the polynomial

$$y(x) = A + Bx + Cx^2 \tag{7.2}$$

for the various data points on the curve are given by the following set of equations:

$$\begin{aligned}
r_0 &= A + Bx_0 + Cx_0^2 - y_0 \equiv y(x_0) - y_0 \\
r_1 &= A + Bx_1 + Cx_1^2 - y_1 \equiv y(x_1) - y_1 \\
r_2 &= A + Bx_2 + Cx_2^2 - y_2 \equiv y(x_2) - y_2 \\
r_3 &= A + Bx_3 + Cx_3^2 - y_3 \equiv y(x_3) - y_3 \\
&\vdots \\
r_n &= A + Bx_n + Cx_n^2 - y_n \equiv y(x_n) - y_n.
\end{aligned} \tag{7.3}$$

The criterion of the least-squares method is that the best fit occurs for a given polynomial when the sum of the squares of the residuals is minimized.

The constants A, B, and C in Equation (7.2) can be determined by utilizing the following procedure:

$$f(A, B, C) = \sum_{i=0}^{n} r_i^2 = \sum_{i=0}^{n} (A + Bx_i + Cx_i^2 - y_i)^2 \tag{7.4}$$

To minimize Equation (7.4), it is necessary that its partial derivatives with respect to A, B, and C are zero. Therefore, the following set of equations result:

$$\sum_{i=0}^{n} 2(A + Bx_i + Cx_i^2 - y_i)\Delta x = 0 \qquad (7.5)$$

$$\sum_{i=0}^{n} 2x_i(A + Bx_i + Cx_i^2 - y_i)\Delta x = 0 \qquad (7.6)$$

$$\sum_{i=0}^{n} 2x_i^2(A + Bx_i + Cx_i^2 - y_i)\Delta x = 0 \qquad (7.7)$$

The solution to these simultaneous equations, using classical algebraic techniques, results in the best estimates for the desired constants.

Harmonic Analysis of Periodic Functions

The digital computer can be a great asset for determining characteristics of functions which are periodic, such as a.c. voltages, by utilizing the techniques of harmonic analysis. As an example, let us consider two functions of the same independent variable which are orthogonal to each other. By orthogonality, we mean that the integral of their product over a specified interval is zero. Considering harmonic functions, orthogonality results in the following set of relationships:

$$\int_{-\pi}^{\pi} \sin mt \sin nt\, dt = 0 \qquad (m \neq n) \qquad (7.8)$$

$$\int_{-\pi}^{\pi} \sin mt \cos nt\, dt = 0 \qquad (7.9)$$

$$\int_{-\pi}^{\pi} \cos mt \cos nt\, dt = 0 \qquad (m \neq n) \qquad (7.10)$$

To find an expression for a periodic function, we will utilize least-squares approximations incorporating orthogonal functions.

Assume that we are determining the harmonic expression for an a.c. periodic voltage of frequency ω. The function can be represented by the following series:

$$e(t) = \frac{A_0}{2} + A_1 \cos \omega t + A_2 \cos 2\omega t + A_3 \cos 3\omega t + \ldots + B_1 \sin \omega t$$
$$+ B_2 \sin 2\omega t + B_3 \sin 3\omega t + \ldots \qquad (7.11)$$

This can also be written as

$$e(t) = \frac{A_0}{2} + \sum_{k=1}^{n} (A_k \cos k\omega t + B_k \sin k\omega t). \quad (7.12)$$

The coefficients A_k and B_k are the Fourier coefficients which are defined as

$$A_k = \frac{1}{\pi} \int_{-\pi}^{\pi} e(t) \cos k\omega t \, d(\omega t) \quad (7.13)$$

$$B_k = \frac{1}{\pi} \int_{-\pi}^{\pi} e(t) \sin k\omega t \, d(\omega t). \quad (7.14)$$

The least-squares criterion for this analysis is that the following expression, which utilizes an n-harmonic term truncation of the series of Equation (7.12), shall be minimized:

$$\int_{-\pi}^{\pi} \left[e(t) - \frac{A_0}{2} - \sum_{k=1}^{n} (A_k \cos k\omega t + B_k \sin k\omega t) \right]^2 dt. \quad (7.15)$$

Determining the A_k and B_k used with the approximating truncated series

$$e(t) = \frac{A_0}{2} + \sum_{k=1}^{n} (A_k \cos k\omega t + B_k \sin k\omega t) \quad (7.16)$$

allows us to approximate the function $e(t)$ as closely as desired by increasing n.

If the form of the function is not known in advance, then a least-squares approximation can be made by solving the normal equations simultaneously for the constants. This technique, however, is very slow and requires the computation of many terms. This implies that many linear equations must be solved, and this is not very efficient utilization of a computer. For this case, the method presented in Reference 5, which reduces the required storage capacity of the computer, should be utilized.

Uncertainties Associated with System Test[6]

Experience has shown that the desired measured value of a parameter is related to its actual value by means of a Gaussian probability density function. Assuming that the probability density function has a standard deviation σ and it is centered at the desired ideal value $\bar{x}$, then the probability that the actual value x lies within a given range is equal to the area under the density curve in that range.

The *probability density function*, $p(x)$, for this situation can be expressed by the following relationships:

$$p(x) = \frac{1}{\sqrt{2\pi}\sigma} \exp\left[-\frac{(x - \bar{x})^2}{2\sigma^2} \right]. \quad (7.17)$$

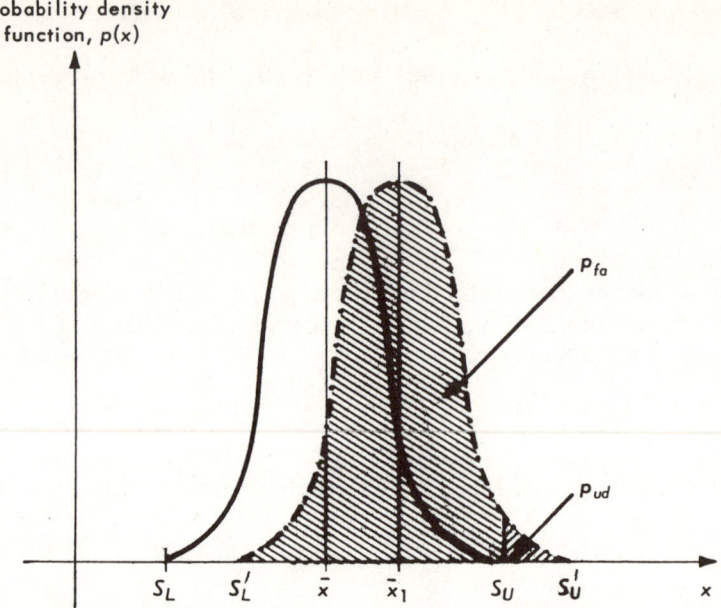

Figure 7.1. Probability Relationships for Measuring an Unknown Value

Figure 7.1 illustrates the relative relationships of this function with the systems lower and upper specification limits, S_L and S_U, respectively. If $\bar{x}$ should change its value to $\bar{x}_1$, then the area under the tail of the curve, which lies beyond S_U, equals the *probability of an undetected defect* p_{ud}. This is illustrated in Figure 7.1 by the dashed curve. The relationship for p_{ud} is given by

$$p_{ud} = \frac{1}{2\sqrt{\pi}\sigma} \int_{S_U}^{S_U=\infty} \exp\left[-\frac{(x-\bar{x}_1)^2}{2\sigma^2}\right] dx. \qquad (7.18)$$

The probability that the measured value falling outside the lower and upper limits is a false alarm equals the probability that the actual value of the measured parameter falls within the specification limits. The *probability of false alarm* p_{fa}, therefore, equals the area under that part of the density function curve that lies within the specification limits, as indicated in Figure 7.1. The expression for p_{fa} in terms of a specific measured value $\bar{x}_1$ is given by

$$p_{fa} = \frac{1}{\sqrt{2\pi}\sigma} \int_{S_L}^{S_U} \exp\left[-\frac{(x-\bar{x}_1)^2}{2\sigma^2}\right] dx. \qquad (7.19)$$

Based upon tolerable values of p_{ud} and p_{fa}, decisions can be made as to the requirements for the accuracy of the system's test equipment. If too small a value for p_{fa} is chosen, the system test equipment accuracy can become excessively high. An important trade-off exists in this area between the probability of having a false alarm occur and the accuracy of the required test equipment.

Example of Testing a Large Naval Shipboard System[6]

The system considered for testing in this section is a large naval shipboard system. The use of a specific system offers several advantages in this presentation. By orienting the discussion toward a specific, very difficult system test problem instead of a very general, abstract one, the reader will gain much more valuable insight.

The reader should attempt to grasp the test concepts presented and to relate them to his own past and present experience, to capture the physical meaning of the concepts presented. The problems and techniques are comparable in many respects with the testing of other large-scale systems. The design of a flexible, centralized, automatic system for on-line testing of a shipboard system begins with estimates of what tests are required and a determination of the quantities of units which must be tested. This must then be compared with what is practical, feasible, and economical.

Figure 7.2 illustrates a functional block diagram which is representative of an automatic testing system.[7] This configuration provides the major functions of generation of stimuli, switching, programming, comparison, evaluation, display/recording. Applying this approach to the automatic testing of complex shipboard systems requires the search of as many malfunctions as possible and the detection of degradations. From a practical viewpoint, it is not possible to test all of the equipment, units, and circuits for degradation and malfunctions. Since checkout of the last few components or circuits in a system becomes exponentially more costly, it is not practical to design an automatic test system to check very reliable, or mechanically and electrically isolated, or unique system elements.

On-line testing complex shipboard systems presents a number of additional unique problems.[8] Since the system must be tested while it is operating, test signals must be conditioned and applied in a manner which does not interfere with equipment operation. Furthermore, the interface problem between the automatic test system and the operational system requires special attention. In practice, it is necessary to integrate the test facility with several operational systems which are usually manufactured by different firms and installed on the ship at different times. As a result, the prime equipment manufacturer usually specifies the test logic, buffer networks, sensors, and programming for the automatic test system.

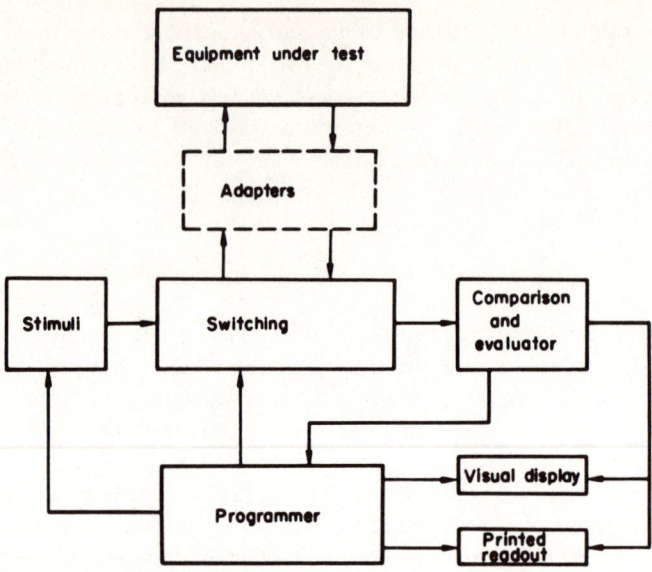

Figure 7.2. Functional Flow Diagram of an Automatic Testing System

A basic goal for automatically testing a multisubsystem shipboard environment containing sensors, missiles, computers, navigation, communication, displays, and command and control systems is to minimize the time-consuming manual actions of performing a test. Deciding whether a circuit, device, or subsystem has failed is basically subjective and presents a problem in on-line testing of complex shipboard systems. Catastrophic failures pose no problem. However, if a piece of equipment has drifted out of operational tolerance, it is necessary to determine which of the many parts may have caused the degradation. Therefore, rather than have a simple **GO**, **NO-GO** indication, it is desirable for the on-line test system to incorporate some indication of degradation. To alleviate the need to rely on manual decisions to determine the values of **GO** and **NO-GO**, another approach is to have the operating equipment set its own limits by periodically checking various test points after knowing that the equipment is operating satisfactorily. Therefore, a high and low limit of acceptable performance can be obtained. After any subsequent performance check on a device, test values outside the stored limits are an indication of a probable malfunction.

Degradation monitoring is very important on complex shipboard systems to alleviate the problem of repairing equipment at a time when it is required for operational use. By monitoring degradation, therefore, correc-

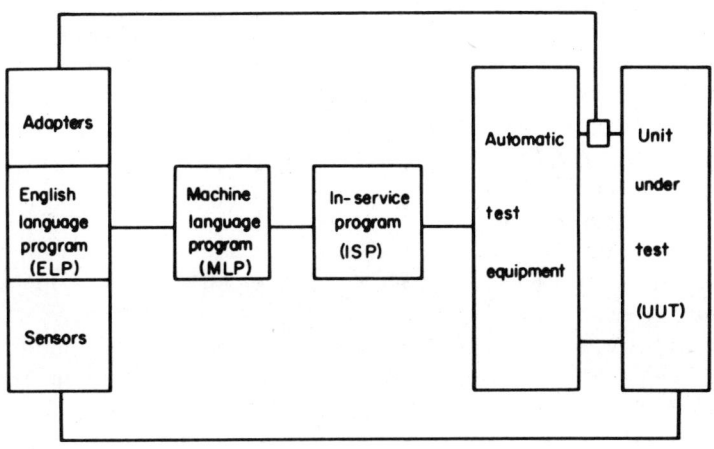

Figure 7.3. Major Features of Automatic Test Software

tive action can be taken before failure occurs or before the equipment has drifted out of tolerable operation.

Software requirements associated with a centralized, on-line, automatic shipboard test system must be carefully considered. The software must function with various ship classes containing different operating systems. In practice, these systems are usually in various states of modification. This presents a difficult problem since the test system must interface with a wide variety of units to be tested, and must maintain the accuracy and currency of programs once the system is installed and operating.

Figure 7.3 illustrates the major elements of software required for a representative shipboard automated test system. English Language Programs (ELP) are the end-to-end monitor, test, and fault isolation procedures that are described in systematic English language statements. Machine Language Programs (MLP) are the ELPs that are integrated with machine control instructions, and then converted into the machine language of an automatic test system. In-Service Programs (ISP) are the MLPs integrated with the particular test system.

The conversion of an ELP to a MLP is quite time-consuming if performed manually; moreover, this conversion provides the opportunity for errors. Therefore, computers are usually utilized for each automatic test system to integrate the ELP coded instructions with pertinent system instructions.

In-Service Programs are completed programs, ready for insertion into the particular automatic test system. Validation of an ISP represents a total acceptance checkout, with the individual ISP loaded into the pertinent automatic test system. The automatic test system is connected to the designated system and validation is performed with all systems energized. Feedback links in Figure 7.3 represent information furnished by the user of the test system to report any program deficiencies, interface problems, and recommendations for extension of the tester's utility.

Configuration management control for the software support of the automatic testing of shipboard systems is very critical. Before an automatic test system can be used aboard a ship, the correct ISP for that ship must be inserted. To accomplish this, the equipment baseline for that particular ship must be precisely known and continually appraised for modification needs. The configuration management functions also require determining which modifications affect individual ISPs, the new or modified logic, and applicable parameters.

This example has indicated the capability for satisfying the software requirements of complex shipboard systems. Any possible deficiencies can be corrected with the built-in adaptive features of validation, on-board surveys, and fleet reporting.

Automatic testing of complex shipboard systems increases the degree of dependability, and is a vital requirement for successful ship missions. Operational availability of naval shipboard systems—containing complexes of sensors, missiles, computers, navigation, communication, and command and control systems—can be assessed via the automated test systems so that the command can judge the operability of the system in its intended mission. The concept for automatic testing of complex shipboard systems presented centers about a centralized, on-line test system which adapts to and supports as many operating systems as possible by means of changes to some of its software.

References

1. Dietz, A.G.H. and Eirich, F.R., eds. *High Speed Testing*, vols. 1 to 5. New York: John Wiley & Sons, Inc., 1960-1965.
2. Machol, R.E., ed. *System Engineering Handbook*. New York: McGraw-Hill Book Company, 1965.
3. Joyce, B.T. and Stockton, E.M. "Computer-Controlled Automatic Testing." *Electro-Technology* 78 (October 1966).
4. Diaz, G. "Computer Methods for Analyzing Test Data." *Electro-Technology* 71 (May 1963).
5. Clenshaw, C.W. "Curve Fitting with a Digital Computer." *Computer Journal* 2 (January 1960).

6. Shinners, Stanley M. "Considerations for Automatic Testing of Complex Shipboard Systems." *Computers and Electrical Engineering* 1 (June 1973) pp. 73-81.
7. Smith, E.F. "A Concept for a Flexible, Automatic Checkout System." *Proceedings of the June 1965 Automatic Support Systems Symposium for Advanced Maintainability*. IEEE St. Louis Section, pp. 4B1-4B10.
8. Margulies, G. "Navy On-Line Testing." *Proceedings of the 1966 Automatic Support Systems Symposium for Advanced Maintainability*. IEEE St. Louis Section, pp. 2E5-2E8.

8
Systems Management

Introduction

Effective management of the systems engineering process is essential for the proper execution of a program. The system can function effectively only in a well-structured hierarchy of authority. Systems engineering management is concerned with monitoring and controlling the activities associated with synthesizing, producing, and testing a coherent system so that it may achieve its intended requirements. The ingredients of systems management are managers, an organization structure, and control/monitoring devices. Systems management is a process which must be exercised throughout all phases of a system's life cycle to ensure the integrity of the system design.

Why is the management of some corporations/divisions better than others? Why is the management of some programs within a given corporation/division better than others? The answer lies in the fact that management is concerned with a very important variable—people. It involves organizing, planning, selecting, controlling, directing, motivating, and evaluating subordinates of the system team.

For the system team to function cohesively, proper channels of direction must be established between the various levels of the team. The key ingredient of this process is communication. Communications must go in both directions and will be effective only if persons with the common goal of solving shared problems have a mutual desire to do so.

Since systems management is concerned with controlling schedule, cost, and human talent to achieve desired system performances and reliability/maintainability/availability (R/M/A), systems managers need clear visibility of the progress of these various factors. In addition, they must depend upon accurate data. The principle systems management control device is the Work Breakdown Structure (WBS), and the Management Information System (MIS) is used for monitoring progress. The WBS and MIS are discussed in the following sections.

Work Breakdown Structure

The WBS is a very important first step which the program manager must develop in order to establish cost and schedule control of the various work packages. The term "work package" refers to a segment of work required to complete a specific aspect of the program in a certain period of time.

The development of the WBS starts with the overall system as a whole, and then works its way down to the lowest identifiable work package level. For example, consider a weapons system containing a tracking radar, missile, launcher, computer/software, and communications subsystem. The weapons system would be at the top level; the subsystems consisting of the tracking radar, missile, computer/software, and communications would all be at the second level. In addition, test and evaluation and training would appear at the second level. Under each of these second-level efforts, work packages at lower levels would be defined. This would continue to the very lowest level of work package which represents an identifiable work effort, such as designing a component, producing a document to test the system, producing operating and maintenance manuals, producing a training manual, etc.

Figure 8.1 illustrates a WBS tree for the weapons system being considered. Note the logical groupings of the functions and tasks depicted. Associated with the WBS tree, a work package dictionary is prepared to describe the inputs from other program tasks, specific work content to be performed, and outputs to other program tasks. The WBS forms the basis for engineering the system. For example, the tasks defined in the tree are used to develop the time-phased program milestone system schedule, program cost, reliability/maintainability/availability, risk analyses, etc.

The time-phased program milestone schedule will indicate the initiation and completion of each task. This schedule would indicate the projected completion dates of all important milestones such as system analysis, simulation, system design, breadboard engineering, development tests, qualification tests, evaluation tests, etc.

The WBS and the time-phased program milestone schedule provide the program manager with visibility for structuring the program effort by depicting when each task should be appropriately accomplished. In addition, it will show what different kinds of engineering skills are required for different phases during the life cycle of the program, and will permit the program manager to select and acquire a proper mix of specialist and generalist skills over the program's duration.

Periodic reviews must be scheduled during the course of the program to assess the status of the effort. Based on these reviews, the amount of manpower applied to the various tasks in the various time periods will be reestimated, and the program schedules and budgets will be adjusted accordingly.

Management Information Systems[1,2,3]

The efficient treatment of management information flow in organizations is

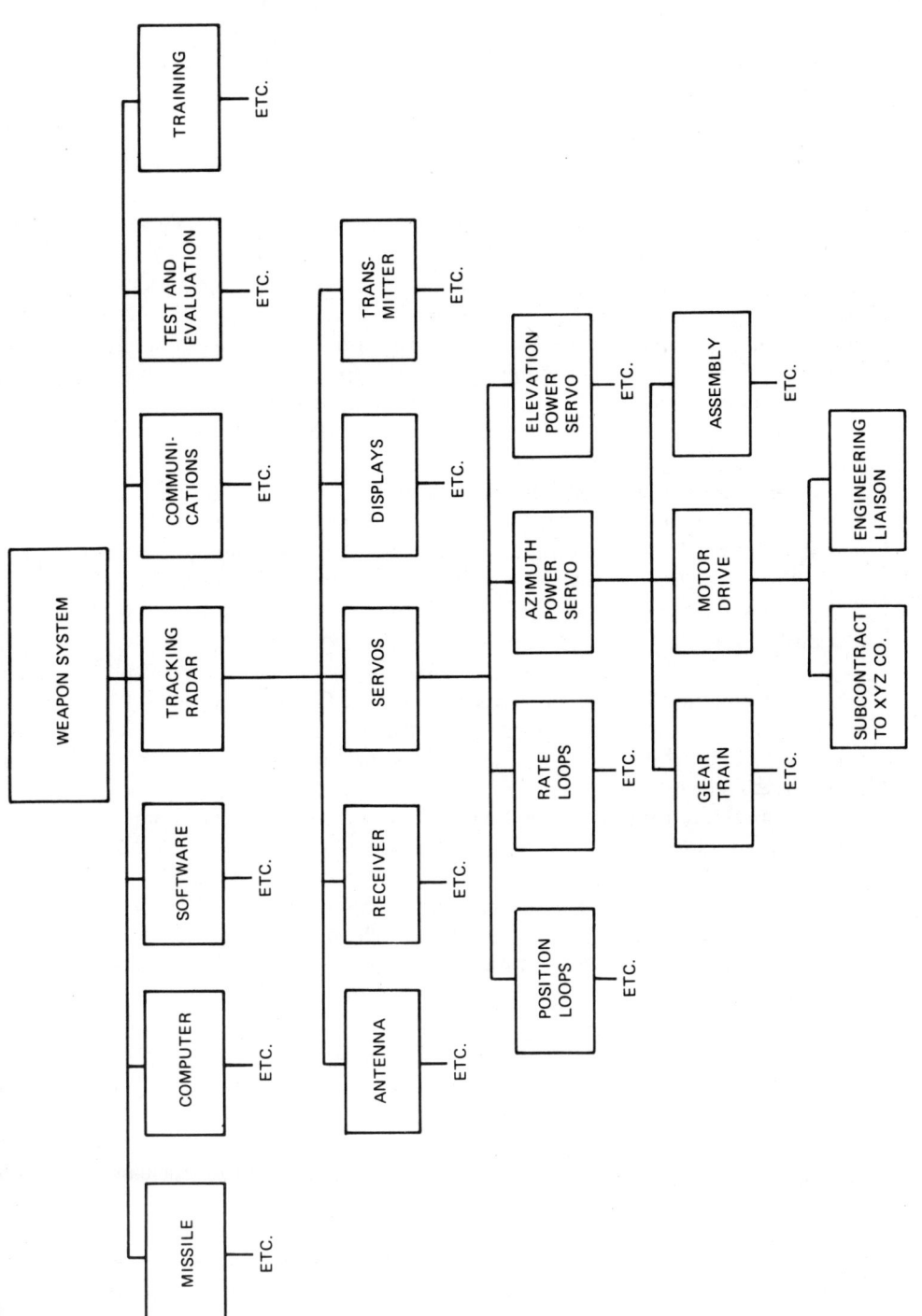

Figure 8.1. A Weapons System Work Breakdown Structure

a very important problem. Information is a very basic resource of each organization, and management must strive to utilize it fully and effectively. Data must be summarized and transformed into many types of measures and statistics which can be utilized for many purposes at many times. Management Information Systems (MIS) collects all kinds of data which can be of use to the organization accurately and efficiently. The system can then use the data to provide comprehensive and meaningful information to all groups and functions.

MIS is inherently linked with the computer because of its data processing power. However, a characteristic of the MIS is recognition of the interaction of man and machine in the total environment of organizational management. For proper functioning, it must provide efficient interrelationships between man and machine so that the total system operates reliably and efficiently. Therefore, an effective MIS provides a two-way communication between computer and people.

A MIS performs the following functions:

1. Supplies complete and timely data which can be used in the decision and planning process by determining the full effect of a decision in advance.

2. Prepares and presents information in a uniform manner and therefore eliminates problems associated with the use of inconsistent and incomplete data.

3. Utilizes specialized mathematical relationships for analyzing data which permits the forecasts of future relationships based on past events.

4. Reports to each management level only the required degree of detail for making decisions quickly and minimizes the time needed for analysis and interpretations.

5. Effectively uses personnel and data processing equipment permitting optimization in speed and accuracy at the lowest cost.

Figure 8.2 illustrates the sequence of events required for the establishment of a new MIS.[3] Emphasis is on the team effort of managers and system engineers required to accomplish this effort. The major phases associated with this chart are deliniated into these four phases: ① Investigation; ② Phase I Design; ③ Phase II Design; ④ Implementation. The content of these phases are as follows:

1. *Investigation*—A task force study team, comprised of representatives from all major activities of the company, is established to perform a survey and investigate the needs and requirements of the MIS. These members should be experienced personnel from the various departments. This group then establishes a project plan concerned with performance, schedule, and time for conceptual design. The team assigns responsibilities and establishes a schedule for completion dates. Needs are determined, and the present system is analyzed. In addition, the team assesses the organization's present reporting system, and analyzes all reports which are

generated. The task force study team determines which reports are redundant and which are not needed, and identifies voids in the reporting process. To accomplish this, information needs are discussed with all levels of management from the viewpoint of present and future requirements.

    2. *Phase I Design*—This phase is concerned with establishing a conceptual design, gross analyses, synthesis, and determining performance specifications for the MIS.

    3. *Phase II Design*—A detailed design is performed in this phase, and the design specifications are reviewed. As part of this phase, a detailed analysis of the organization's present data processing equipment is needed in order to determine adequacy or inadequacy.

    4. *Implementation*—Installation and construction of the system begins during this phase.

An effective MIS must supply only required data to each level of control. This data must be presented in a manner which permits understanding for quick and decisive action. In addition, it should provide a method for measuring the effectiveness of the action that has been and is being taken. To accomplish this, the MIS should provide the following four primary types of reports:

    1. *Monitoring reports* are the periodic performance reports usually prepared for management to provide an overview of the whole performance of the activity. Typical monitoring reports are project progress reports, end-of-month sales reports, experience reports, etc. Actual and plan figures are indicated, with significant differences denoted. If additional detail between actual and plan is required, the demand report is utilized.

    2. *Demand reports* are the features of on-line query and response to allow management to probe systematically for more detail when analyzing a problem. Illustrations of demand reports are costs-per-sales-calls, number of orders due for shipment but not shipped, machine utilization analysis, etc. These reports can be used to determine deviations between actual and plan as focused in the monitoring reports. In addition, they can be used to determine the causes of triggered reports.

    3. *Triggered reports* are prepared if the deviation from the plan is significant and requires an immediate corrective action. Triggered reports cover operations involving quality control, machine downtime control, inventory control, etc. These reports are produced only when an out-of-control situation is caused by the MIS.

    4. *Planning reports* are used to plan the needs of proposed decisions. As an illustration, if management proposes to increase sales by 20 percent in the new fiscal year, then more salesmen, advertising, secretaries, office space, telephones, etc., are required. Based on past experience, requirements and activities can be logically developed to establish such action. In

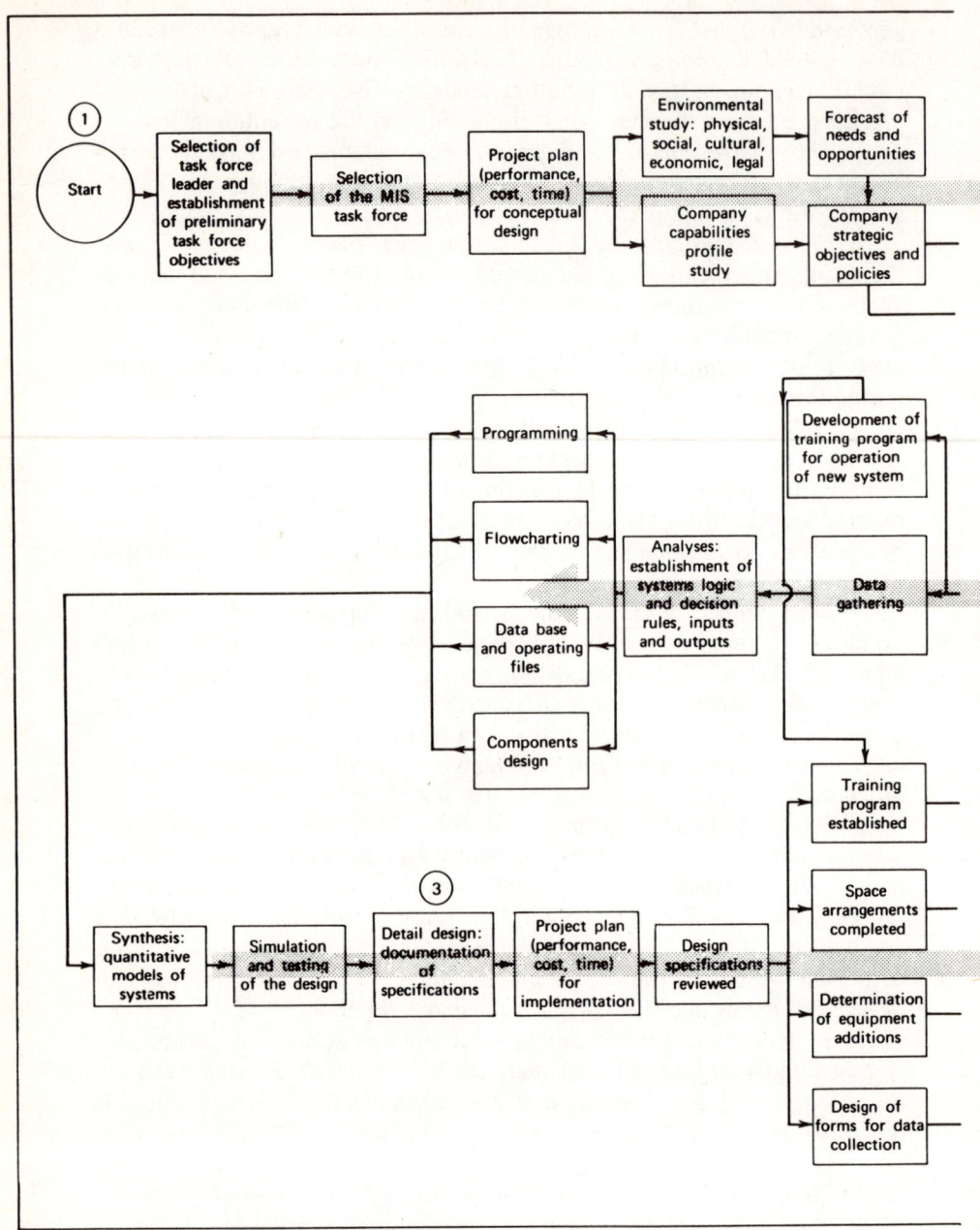

Source: Reprinted by permission of John Wiley & Sons, Inc. from *System Analysis Techniques*, edited by J. Daniel Couger and Robert W. Knapp, 1974, pp. 90-91.

Figure 8.2. The Development Process of an MIS

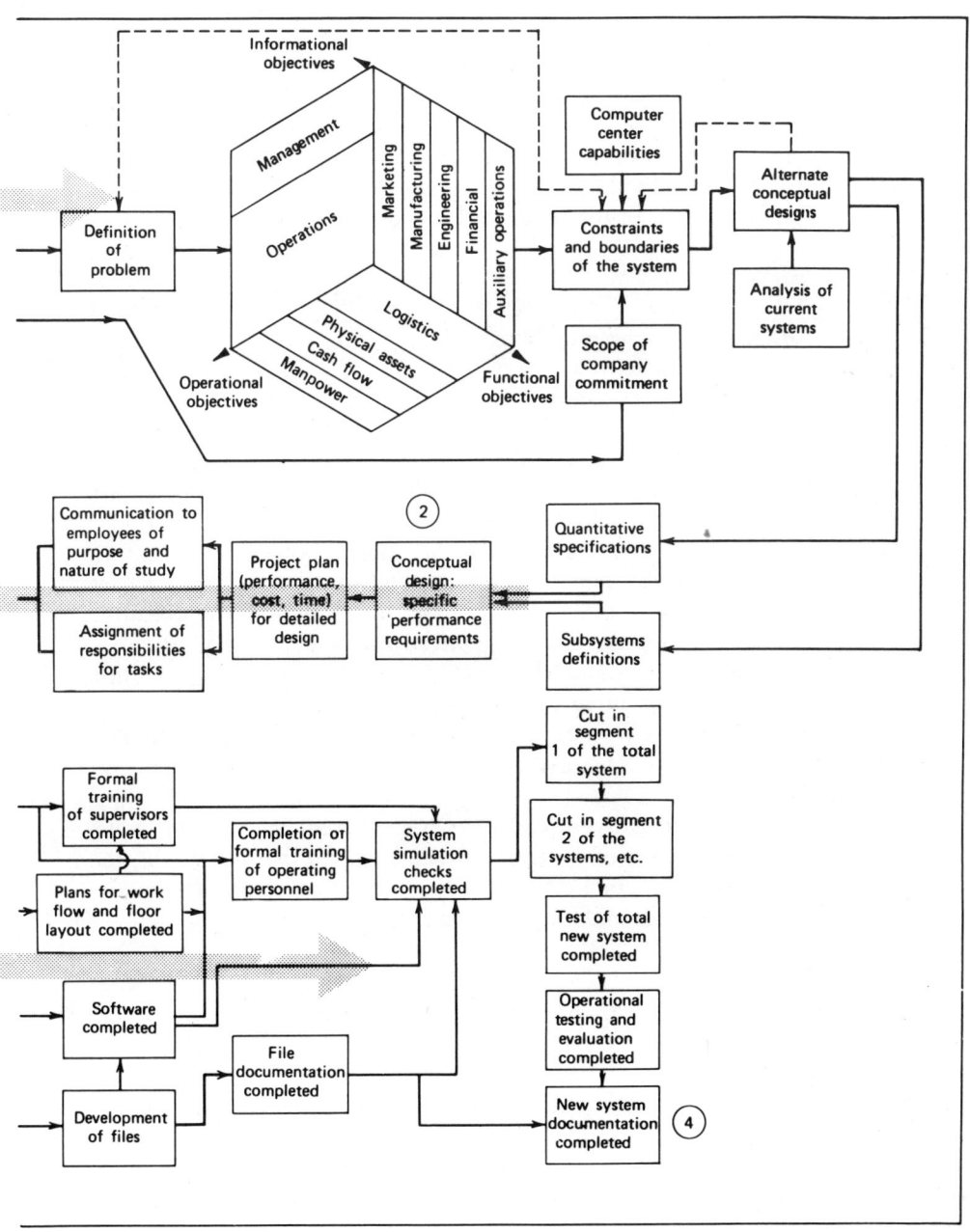

addition, the planning reports contain the capability to test proposed actions by utilizing the simulation capability of the computer.

The awareness of management that computers can be utilized to produce these meaningful reports on a timely basis, has provided the impetus to the formation of MIS groups within organizations. With the additional information provided by the MIS to management, they are able to plan and control the operations of the organization more effectively.

Systems Management Considerations[4]

Systems management is concerned with the problem of integrating the often conflicting effects of performance, reliability/maintainability/availability, schedule, and cost. The technique for accomplishing this is the appropriate consideration of assumptions, alternatives, and risks which are inherent in any possible selection.

An *assumption* deduces the existence of a fact, although it is not known with certainty that it is a fact based on the known data. A manager uses assumptions to deal with difficult problems whose solutions are obscure.

Alternatives provide the manager with several acceptable solutions to consider. The evaluation of alternatives for selecting the optimum solution is usually difficult in practice since the decision is seldom clear.

Risk represents a measure for system failure. Low statistical probability characterizes high risk. However, exact measure of risk are not always quantifiable.

Assumptions, alternatives, and risks are inherent ingredients in the decision-making process of systems management. In the real world, the selection of a particular course of action based on often-conflicting factors is a process of trade-off and compromise.

Assumptions

Assumptions are not demonstrable; therefore, big unknowns often exist. For this reason, it is desirable in systems management to make as few assumptions as possible since each assumption tends to weaken the resulting decision and action.

If an assumption is introduced as a convenience for reducing risk, then the systems manager is faced with a potential major problem. A poor managerial decision may result because the risk may not have been weighed adequately. When there are more assumptions than statements of fact, risk is greatest and the resulting managerial decisions are really guesses.

The program manager should list the alternative risks in an ordered manner from greatest to least. For each category of risk, he should have a rationale and a number of assumptions. The program manager should attempt to eliminate the alternatives containing only high risk and focus attention on the alternatives with moderate and low risk. For these alternatives, he should carefully analyze the assumptions. In practice, he may be able to eliminate some assumptions but may be forced to add others. Careful analysis may permit the program manager to balance assumptions and risks.

Alternatives

Alternative solutions permit the program manager a range of choices. Each alternative solution, which must be viable, has advantages and disadvantages.

Alternative solutions are evaluated on the basis of performance, reliability/maintainability/availability, and schedule and cost factors. The program manager must select the alternative solution that best serves the customer and the organization from various, but often conflicting, viewpoints.

In general, alternatives can be classified as functionally or operationally different.[4] Functional alternatives differ with respect to the method of accomplishing a task. For an example of functionally variant alternatives, consider a private automoblie and a bus, both of which are alternative modes of transportation for accomplishing the same goal. Operational alternatives differ with respect to the operation of the system. Operationally variant alternatives, for example, might be different designs of a bicycle (single-speed, three-speed, and ten-speed) which accomplish the same goal.

Very often a simple matrix representation may aid in visualizing the problem. For example, consider the design of a high-performance fighter aircraft. Table 8.1 illustrates the comparison of three alternative designs. Alternative A has the greatest performance, highest R/M/A, will take the longest to develop, and will cost the most. At the other end of the spectrum, Alternative C has the least performance, least R/M/A, will take the shortest time to develop, and will cost the least. Alternative B is a compromise in every respect between alternatives A and C, and is the preferred alternative from a practical viewpoint in this simple academic example. In the real world, the alternatives are never this clearly delineated, and a selection is never this easy.

Table 8.1
Alternative Designs of a High-Performance Fighter Aircraft

Design \ Parameter	Performance	R/M/A	Schedule	Cost
A	GREATEST	HIGHEST	LONGEST	HIGHEST
B	MEDIUM	MODERATE	MIDDLE	MEDIUM
C	LEAST	LEAST	LEAST	LEAST

Table 8.2
Selection of Alternative High-Performance Fighter Aircraft Designs with the Dimension of Risk Added

Design \ Parameter	Performance	R/M/A	Schedule	Cost	Risk
A	GREATEST	HIGHEST	LARGEST	HIGHEST	LEAST
B	MEDIUM	MODERATE	MIDDLE	MEDIUM	MIDDLE
C	LEAST	LEAST	LEAST	LEAST	GREATEST

Risk Assessment

In evaluating the various alternative solutions, the risk of each one must also be considered. Let us add this dimension to the high-performance fighter aircraft design example considered in Table 8.1. In the resulting matrix of Table 8.2, risk has been shown for the three alternative designs. The risks shown for alternatives A, B, and C have been purposely chosen to keep this theoretical problem simple. For this academic problem, alternative B would still be the desired alternative. On the other hand, if alternative B had the greatest risk and alternative C had the least risk, then a good case could be prepared for utilizing alternative C instead of alternative B.

Risk assessment is a very important factor to be considered by the program manager at the beginning of a program for evaluating the possible losses involved in executing a program. Overall program risk is based on these major factors: theoretical feasibility, conciseness of program definition, performance requirements, reliability/maintainability/availability, schedule, and cost.

Consider these risk analysis factors individually:

Theoretical feasibility is concerned with whether there is theoretical evidence to support the approach. To make a proper risk assessment, the

Table 8.3
A Task Risk Analysis Overview List: Weapons System Program, Tracking Radar Task

	Risk Level		
Appraisal Factor	Low	Moderate	High
Theoretical Feasibility	X		
Consciences of Program Definition	X		
Performance Requirements		X	
Reliability/Maintainability/Availability			X
Schedule	X		
Cost		X	

Overall Risk Percentage: 20.

available theoretical literature should be reviewed for evidence that the approach is feasible. Another supplementary approach is to call upon expert consultants in the field at the very beginning of the effort.

Conciseness of program definition refers to how well the program has been defined in terms of tasks to be performed. Is the customer clear in his own mind what he wants, or are you being forced into accepting an ill-defined, open-ended contract with much risk attached?

Performance requirements are concerned with the risk involved in attaining the system's technical requirements. The question to be answered focuses upon the ability to meet the performance requirements with the resources available based on the organization's prior experience with similar programs.

Schedule risk can be answered in a straightforward manner by reflecting on one's past experience with similar programs. Based on this background, what are the risks in meeting the schedule?

Reliability/Maintainability/Availability (R/M/A) risk must be considered on a component, subassembly, assembly, subsystem, and system basis. Do the models indicate that the specified R/M/A can be met? What R/M/A must be demonstrated? What is the experience of the organization and others on similar systems? Answers to these questions will aid in formulating an answer to the degree of risk in the R/M/A area.

Cost risk analysis has several aspects. First, is your cost estimate well-founded? Second, do you believe that your cost is in line with your competition? Third, does the customer have the funds to buy what you're estimating? These questions must be answered in order to formulate an answer to the cost risk involved.

To focus attention on these various factors on a task basis, the Task Risk Analysis Overview List, shown in Table 8.3, is utilized. One of these

Table 8.4
Guidelines for Assigning Risk Percentage

Risk Percentage Guideline	Description
0-10	Existing operational hardware exists, but some minor modifications are required.
10-30	Approach is technically feasible and demonstration breadboards exist; equipment has never been operational before.
30-60	Approach appears technically feasible, but certain aspects of the design may be pushing the state of the art.
60-100	Theoretical analysis is complete and approach appears feasible, but the requirements are considerably beyond existing hardware and the state of the art.
100-200	Comparable technology does not exist.

lists should be prepared for each major task of the program. In the example of Table 8.3, attention is focused on the design of the tracking radar which is part of the weapons system considered previously in this chapter.

The risk percentage determined for the task, and listed at the bottom of Table 8.3, is not based on any specific formula, but on the careful judgment of an experienced program manager. Guidelines based on past program history can be used to assist in this determination. Table 8.4 illustrates a set of possible guidelines which can be used to assign risk percentage. This table shows gross ranges for assigning risk; the actual value selected is the final judgment of an experienced manager.

The various Task Risk Analysis Overview Lists can be summarized on the Work Breakdown Risk Analysis Summary Sheet shown in Table 8.5. An example for developing the servo-driven tracking radar previously considered is illustrated. In this example, it is concluded that a risk of 20 percent exists in developing the $4.0M tracking radar, or a risk of $800K.

Very often the risk analysis should not be viewed strictly for the program currently considered. For example, will the winning of this program result in follow-up business which will provide additional business/profits, and reduce the initial risks involved?

To determine the minimum basic profit resulting if all the anticipated risk occurs, and to estimate the overall real profit which will accrue if follow-on work evolves, let us formulate appropriate expressions to be evaluated. The basic minimum profit, BMP, which may result if all of the anticipated risk occurs, and must be fully absorbed, is given by:

$$\text{BMP} = \frac{PV - R}{V} \; 100\% \qquad (8.1)$$

Table 8.5
A Work Breakdown Risk Analysis Summary Sheet for Weapons System Program

Work Breakdown Level					Description	Cost Estimates	Cost Risk	
I	II	III	IV	V		$M	%	$M
X					Weapons System	40.0	5	2.00
	X				Tracking Radar	4.0	20	0.80
		X			Antenna	0.5	10	0.05
		X			Receiver	0.5	30	0.15
		X			Transmitter	0.8	10	0.08
		X			Servos	1.0	25	0.25
		X			Displays	0.3	20	0.06
		X			Vehicle	0.1	10	0.01
		X			Data	0.5	10	0.05
		X			Training	0.3	5	0.15

where P = profit of basic contract in percent; V = contract value in dollars; and R = risk in dollars. The overall real profit, ORP, to be realized if follow-on business results is given by:

$$\text{ORP} = \frac{(PV - R) + \sum_{i=0}^{n} V_{A_i} P_i R_i}{V + \sum_{i=0}^{n} V_{A_i} R_i} \ 100\% \quad (8.2)$$

where V_{A_i} = value of follow-on contract i in dollars; P_i = profit of follow-on contract i in percent; R_i = probability of being awarded follow-on contract i; and n = total number of follow-on contracts possible.

To illustrate this point, consider the overall weapons system example introduced earlier in this chapter. Suppose that by winning this weapons system, a follow-on spare components contract is assured. Let us assume that the weapons systems contract is for $40.0M with a 10% fixed fee, and has a $2.0M risk. The spares contract is valued at $6.0M, has an associated profit of 20%, and has 100% probability of being awarded. From Equation (8.1), the basic anticipated minimum profit based on a $2.0M overrun (risk) is given by:

$$\text{BMP} = \frac{(10\%)(\$40.0M) - \$2.0M}{\$40.0M} \ 100\% = 5\%. \quad (8.3)$$

From Equation (8.2), the real profit to be realized with the award of the spares contract of $6.0M, having a 20% profit and a 100% probability of being awarded is given by:

$$\text{ORP} = \frac{[(10\%)(\$40.0M) - \$2.0M] + (20\%)(\$6.0M)(1)}{\$40.0 + \$6.0(1)} \; 100\% = 6.96\%. \quad (8.4)$$

As a second example, assume that by winning this weapons system contract, there are two follow-on contract awards possible in addition to the spares contract. There is a 75% probability of being awarded a contract valued at $10.0M having an 18% profit to provide support equipment, and a 50% probability of being awarded a contract valued at $6.0M having a 14% profit to retrofit some equipment. Inserting these values into Equation (8.2), the overall real profit, ORP, of the contract is found to be:

$$\text{ORP} = \left[\frac{[(10\%)(\$40.0M) - \$2.0M] + 20\%(\$6.0M)(1)}{\$40.0M + \$6.0M(1) + \$10.0M(0.75) + \$6.0M(0.5)} \right.$$
$$\left. + \frac{18\%(\$10.0M)(0.75) + 14\%(\$6.0M)(0.5)}{\$40.0M + \$6.0M(1) + \$10.0M(0.75) + \$6.0M(0.5)} \right] 100\%$$

$$\text{ORP} = 8.79\%. \quad (8.5)$$

Therefore, we conclude that the weighted contract values and profit resulting from the follow-on contracts can increase the basic minimum profit from 5% to 8.79%.

The collection of data for a credible risk assessment is extremely important.[5] Intelligent judgments must be exercised at all levels of the systems engineering process. Since the program manager cannot be an expert at all levels of the system, he must depend on the engineers working at the various subsystem and unit levels for assistance in establishing risk. This data must then be filtered and integrated into the overall risk assessment of the system.

To give credibility to the risk assigned to the various units, subsystems, and systems, management has found that a most effective approach is to use a programmed checklist[5] in a structured interview of the engineers involved. Table 8.6 illustrates a typical checklist useful for interviewing engineers during the initial steps of a system's design. This raw data must then be summarized and converted to system risk assessment as illustrated in Tables 8.3 and 8.5.

Communication Aspects of Systems Management

Several aspects of systems management have been analyzed in this chapter by considering techniques used for controlling (WBS), monitoring (MIS),

Table 8.6
Possible Checklist to be Used for Collecting Data For Determining Risk Assessment

RISK ANALYSIS CHECKLIST

PROGRAM _____

TASK _____

1. *Technology*

    What of the following will be required during development for meeting the specified technological requirements:

    ☐ Simulation ☐ Engineering Model
    ☐ Breadboards ☐ Production Model
    ☐ Prototype ☐ Environmental Test Chamber
    ☐ Special Test Facilities

2. *Performance*

    How does the performance required compare with the state-of-the-art:

    ☐ Within existing technology
    ☐ Requires moderate advance
    ☐ Requires major advance

3. *Reliability/Maintainability/Availability*

    How do the R/M/A requirements compare with the state-of-the-art:

    ☐ Within existing capabilities
    ☐ Possible, but never has been demonstrated
    ☐ Beyond existing capabilities

    Are any special demonstrations required to satisfy R/M/A?

    ☐ Yes
    ☐ No

4. *Schedule*

    The number of months required to perform this task is:

    Shortest _____
    Most Likely _____
    Longest _____

5. *Cost*

    a) The estimated cost to accomplish this task is _____ .

    b) Is this estimate based contingent upon the continuation or award of another contract?

    ☐ Yes   ☐ No

    c) What capital equipment is required for this task? _____

Table 8.6 (cont.)

---

6. *Human Resources*

   The qualified and experienced manpower required to perform this task are:

   ☐ Immediately available

   ☐ Available if released from other programs

   ☐ Not available and would have to be recruited from the outside

   Are special consultants required to accomplish this task?

   ☐ Yes   ☐ No

   Name _____ Title _____ Date _____

---

decision making, and analysis (risk assessment). Unfortunately, management's task of implementing these techniques to govern, coordinate, and control the systems engineering process is further complicated since it is necessary to deal with many individuals and intangibles. Unless clear orders, instructions, rules, and guidelines are carefully established and communicated to the systems engineering team, difficulties will arise. Clear and effective communications must also be established between the various decision-making levels. For this to be accomplished, there must be a mutual understanding of goals and viewpoints between the individuals at each decision-making level.

To create and produce an effective system, there must also be mutual communication and understanding between the producer and the customer.[5] Lack of this could cause undercontrol by the customer which would probably result in an unsatisfactory initial system requiring numerous modifications to fulfill the customer's inner expectations. On the other hand, customer overcontrol could also be costly, since it would tend to inhibit creative and inventive talent when they conceive and design the system. Therefore, there must also be effective communications between the customer and the producer.

Effective systems engineering management must be exercised by capable systems-oriented managers working directly with the line team and customer and answering decision-demanding questions on a timely, person-to-person basis. The larger the system involved, the more demanding is the systems management problem and the more difficult is the communications problem. Therefore, the program manager must rely on and make proper utilization of the management techniques presented in this chapter, and effectively communicate with the line team and the customer.

References

1. Lazzaro, Victor, ed. *Systems and Procedures: A Handbook for Business and Industry*. 2nd ed. Englewood Cliffs, N.J.: Prentice-Hall, Inc., 1968.
2. Matthews, D.Q. *The Design of the Management Information System*. New York: Auerbach Publishers, Inc., 1971.
3. Couger, J. Daniel and Knapp, Robert W., eds. *System Analysis Techniques*, New York: John Wiley & Sons, Inc., 1974.
4. Optner, Stanford L. *Systems Analysis for Business and Industrial Problem Solving*. Englewood Cliffs, N.J.: Prentice-Hall, Inc., 1965.
5. Chase, Wilton P. *Management of System Engineering*. New York: John Wiley & Sons, Inc., 1974.

9 Problems

Chapter 1

1.1. Consider a program you are familiar with. Draw a functional diagram illustrating the approximate weighting factors assigned to performance, reliability, schedule, cost, maintainability, life expectancy, power consumption, and weight. How could this have been changed to improve the overall result?

1.2. The various factors considered for systems optimization differs markedly for different programs. Consider the development of the following products:

a. Home movie projector
b. Transistor radio
c. Communications set for infantry troops
d. Control jet for a space vehicle
e. Battery for a submarine

For each of these programs, list the following considerations in descending order of importance for the development of each product if you were the company president:

1. Performance
2. Reliability
3. Schedule
4. Cost
5. Maintainability
6. Life expectancy
7. Power consumption
8. Weight

1.3. In your college career, list your goals in descending order of importance. Do the same for your professional career. What weighting factors would you attach to each?

1.4. A company manufactures electric light bulbs. The main criteria of success are cost, schedule, life expectancy, and power consumption. Draw

---
References within problems are found with the appropriate chapter references.

a block diagram representation illustrating the feedback interrelationships of these factors.

1.5. An electric utility company provides electric power to a rural farming community. The primary considerations of success are cost, maintainability, and power consumption. Draw a block diagram representation illustrating the feedback interrelationship of these factors.

1.6. An airplane manufacturer is developing a modern fighter for an attack aircraft carrier. The primary criteria for success are performance, reliability, schedule, cost, power consumption, and weight. Draw a block diagram representation illustrating the feedback interrelationship of these factors.

1.7. The book *Wealth of Nations* by Adam Smith, published in the eighteenth century, discussed the issue of free competition between the members of an economy. Smith suggested that workers compare various possible employment opportunities and enter the one that offers the greatest rewards. He also pointed out that the rewards diminished as the number of competing workers in a trade rises. Essentially, Adam Smith utilized social feedback mechanisms to explain his theories.[8] Draw a feedback system to represent this process in which r = total rewards averaged over all trades; u = influx of workers into a specific trade; and c = total rewards in a particular trade.

1.8. Figure 9.1 represents the general concept of a command and control system which is essentially a closed-loop system having the capability of changing an environment based on information presented to a central human controller. The decisions are based on information obtained by sensors, communicated by a data link to a data processor, and appropriately displayed to a commander who then closes the loop by communicating commands to the personnel and equipment at his disposal in order to change the overall environment. A complete command and control system includes all subsystems, related facilities, equipment, material, services, and personnel required to operate the self-sufficient system. From a functional systems engineering viewpoint, a command and control system includes sensors, communications, data processing, displays, and the man-machine interface. Synthesize command and control systems for the following applications and discuss the characteristics of the data processing and communications subsystems:

1. An airline reservations system for a trunk carrier that has one central terminal, five major terminals, and six minor terminals.
2. A worldwide satellite communications system containing two major land stations and one major shipborne station.

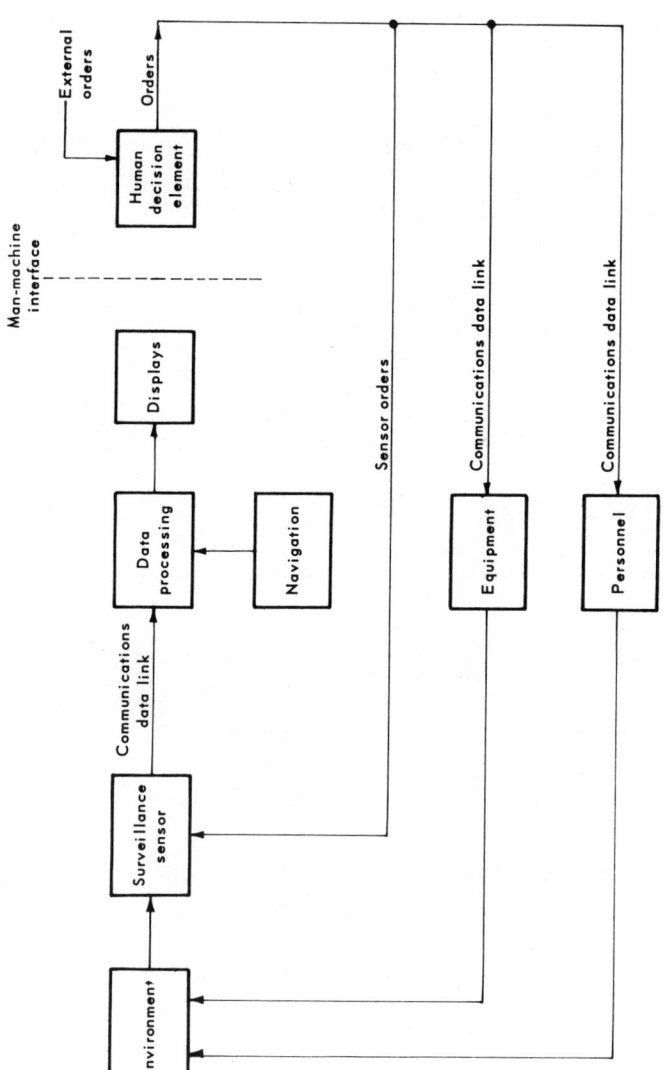

Figure 9.1. General Concept of a Closed-Loop Command and Control System

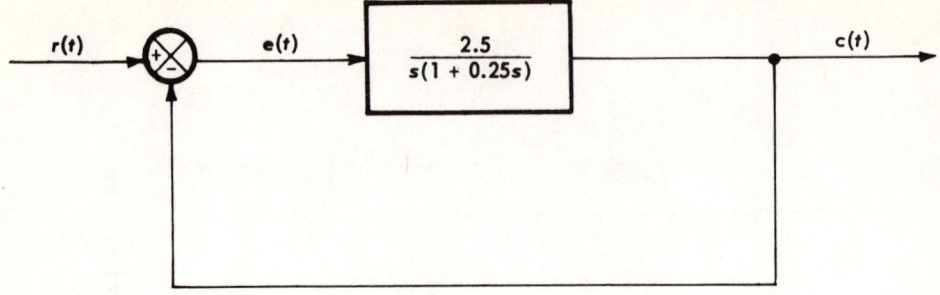

Figure 9.2. Block Diagram of a Linear Control System

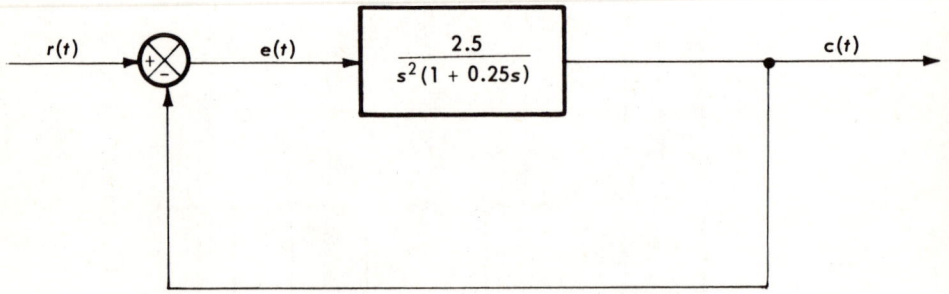

Figure 9.3. Block Diagram for a Control System

3. A strategic command and control system responsible for the defense of the northeastern section of the United States.
4. An antisubmarine warfare system aboard a destroyer.
5. A ground traffic control system for a city of 4 million population.
6. A police command system controlling the operation of twenty stationary posts and forty mobile units from one central headquarters station.

Chapter 2

2.1. The block diagram of a linear control system is shown in Figure 9.2.

1. Determine the steady-state error resulting from a velocity input which can be represented by:

$$r(t) = 10t. \tag{9.1}$$

2. Determine the steady state error resulting from the following input:
$$r(t) = 4 + 6t + 10t^2. \tag{9.2}$$

2.2. Determine the following for the system shown in Figure 9.3:
1. Steady-state error resulting from an input given by: $r(t) = 10t$.
2. Steady-state error resulting from an input given by: $r(t) = 4 + 6t + 3t^2$.
3. Steady-state error resulting from an input given by: $r(t) = 4 + 6t + 3t^2 + 10t^3$.

2.3. Determine the circular probable error for a navigation system whose "along-course" and "cross-course" errors are each 0.5 nautical miles.

2.4. Determine the circular probable error for a navigation system whose "along-course" error is 2 nautical miles and whose "cross-course" error is 0.7 nautical miles.

2.5. Repeat Problem 2.4 with the "cross-course" error doubled.

2.6. An instrumentation tracking radar located in a radome has the following sources of random error:
1. Servo noise—0.02°
2. Receiver thermal noise jitter—0.03°.
3. Data unit readout—0.015°.
4. Data unit noise—0.001°.

Determine the overall RMS error.

2.7. Repeat Problem 2.6 with the servo noise reduced to 0.01°. What conclusions can you reach from your result?

2.8. Repeat Problem 2.6 with the servo noise doubled. What conclusions can you reach from your result?

2.9. If a radome is not part of the radar in Problem 2.6, the following additional random errors caused by the wind are present:
1. Structural—0.015°.
2. Servo—0.04°

What is the resulting overall RMS error? What conclusions can you reach from your result?

Figure 9.4. Tokyo-to-Hakata Super Express Train (Courtesy of Japanese National Railways)

2.10. Railroads are utilizing automatic control systems to an ever-increasing degree.[14] The Tokyo-to-Hakata railroad system, commonly referred to as the new Tokaido line, is a very widely acclaimed example of a modern high-speed rail transportation system. Figure 9.4 is a photograph of one of its trains; Figure 9.5 illustrates an equivalent block diagram for the automatic braking system used to regulate this class of high-speed trains.

1. Determine the steady-state error to an input command of a unit step.
2. Repeat part 1. for a unit ramp input.
3. Repeat part 1. for a unit parabolic input.

2.11. The control of the depth of a submarine illustrated in Figure 9.5 is an interesting system control problem. The goal of this system is to adjust the actual depth C to equal the desired depth R. A pressure transducer measures the actual depth. Any difference between the actual and desired depths is amplified and drives the stern plane actuator through an appropriate angle Θ until the actual and desired depths are equal. An equivalent block diagram is shown in Figure 9.7.

1. Determine the error of this depth control system for an input command of a unit step.
2. Repeat part 1. for a unit ramp input.
3. Repeat part 1. for a unit parabolic input.

Chapter 3

3.1. An electronic system has a MTBF of 1,000 hours and a MTTR of 100 hours. What is its availability?

3.2. Assume that a computer is composed of 1,000 elements, each having a MTBF of 100,000 hours. What is the probability of failure of the system if its cumulative operating time is 10 hours?

3.3. An electronic device can be designed utilizing vacuum tubes or transistors. If tubes are employed, 5 are required, each having an MTBF of 10,000 hours. If transistors are employed, 20 are required and each has an MTBF of 80,000 hours. Does it pay to transistorize the unit from a reliability viewpoint? What is the reliability improvement factor?

3.4. Repeat Problem 3.3 with 40 transistors required. What conclusions can you reach from your results?

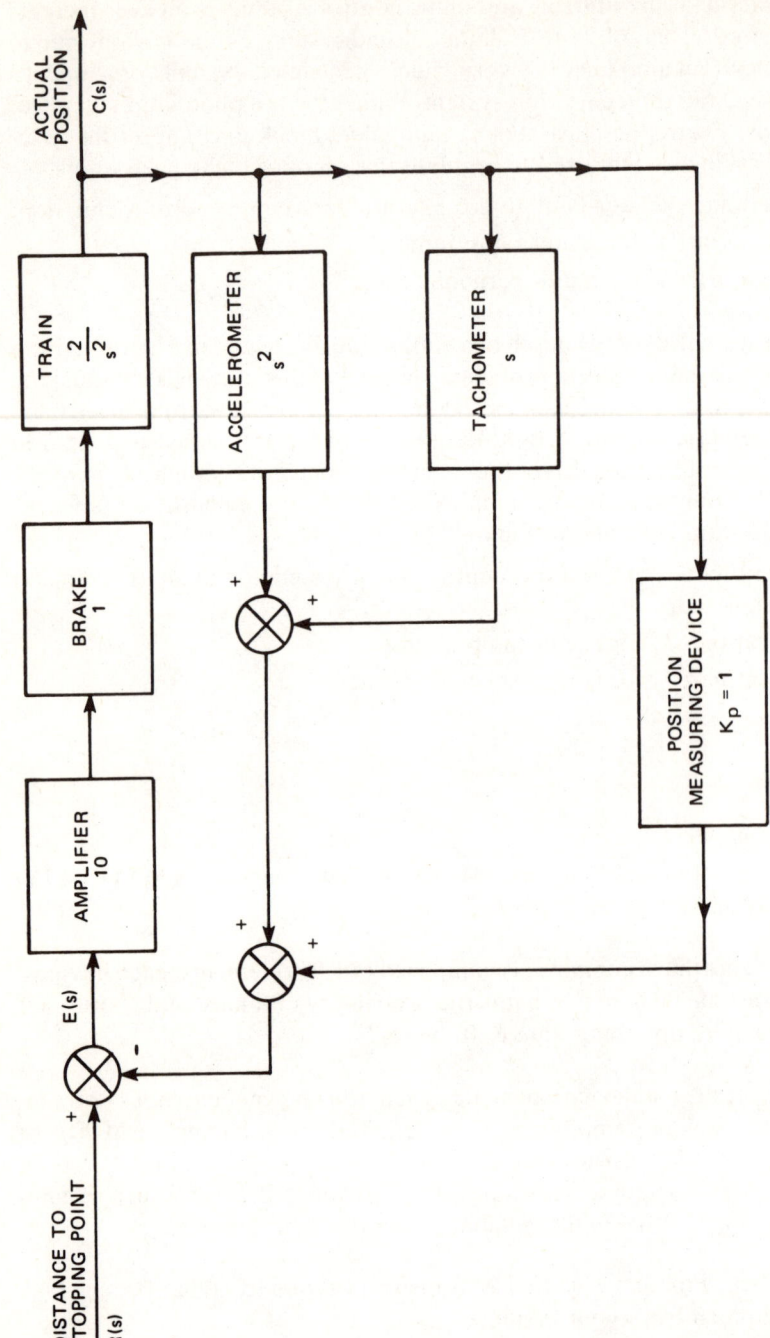

Figure 9.5. Block Diagram for Automatic Braking System of High-Speed Train

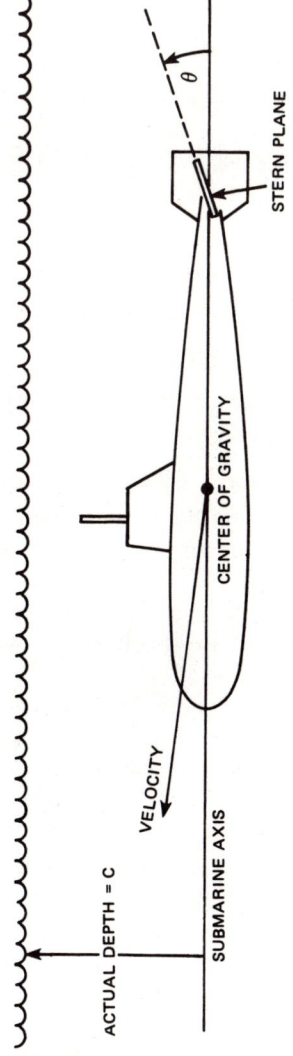

Figure 9.6. Submarine Depth Control Problem

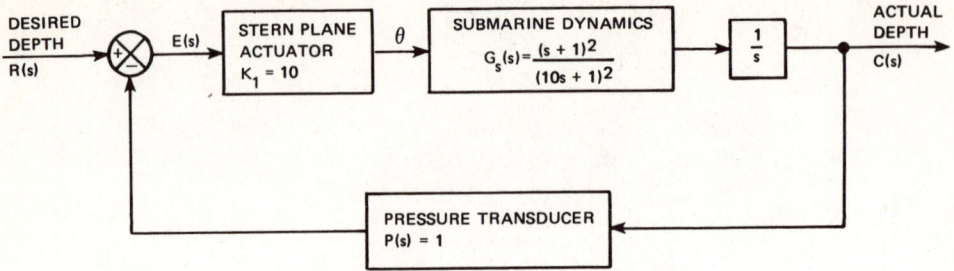

Figure 9.7. Equivalent Block Diagram for Depth Control System of a Submarine

3.5. Repeat Problem 3.3 with 10 transistors required. What conclusions can you reach from your results?

3.6. A military digital computer consists of 5,000 elements, each having an MTBF of 100,000 hours. Its cumulative operating time is 1 hour.
1. What is the probability of failure?
2. What is its probability of failure, if it is designed in a group redundant manner, with each group consisting of 10 elements?
3. What is the reliability improvement factor?

3.7. Repeat Problem 3.6 with each group containing 100, or 500, or 1,000 elements. What conclusions can you reach from your results?

3.8. Consider the digital computer described in Problem 3.6. Consider a design based on partial redundancy where 1,000 of the elements are non-redundant and the remaining 4,000 elements have redundancy. What is the reliability improvement factor?

3.9. Two computer manufacturers are attempting to meet a 99.4% availability guarantee over a six-month test period based on the availability of a 24 hour/day repair service.
1. Computer X is a reasonably reliable system with failures occurring approximately once every five weeks. The manufacturers provide on-call service from the nearest service center during the guarantee period. Due to the location of the center, it takes about five hours on the average to correct each failure. This includes travel time, diagnostic time, and repair time. Does computer X satisfy the availability guarantee?
2. Computer Y has a true MTBF of only 66 hours. What would be the maximum downtime (MTTR) required of Computer Y to satisfy the availability requirements?

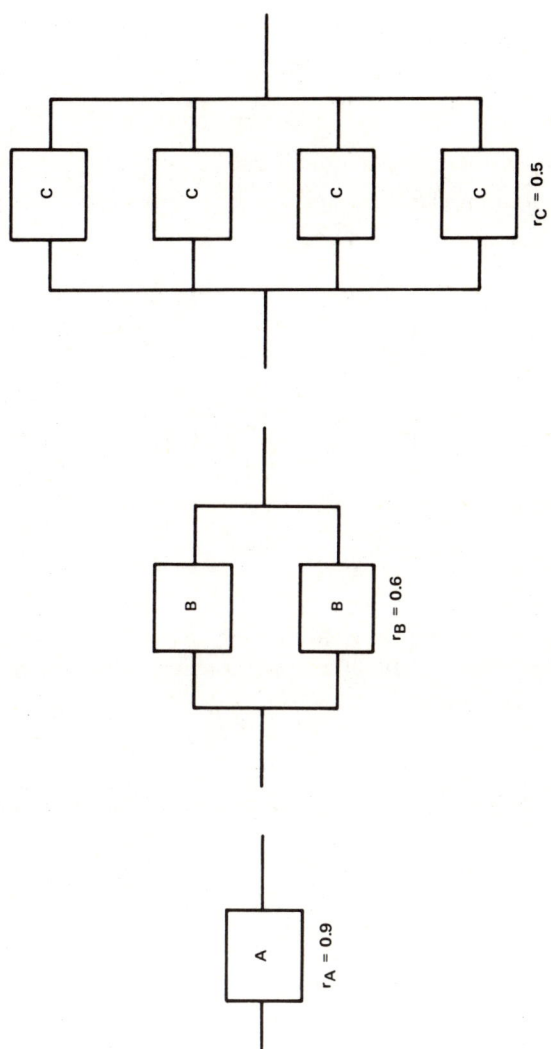

Figure 9.8. Three Computing Systems to Compare for Reliability

3.10. Two units of a system, A and B, have reliabilities of 0.95 and 0.50, respectively. Determine the reliability for the following configurations:
1. A and B are cascaded.
2. A and B are cascaded and each is redundant.
3. A and B are cascaded and only B is redundant.
4. A and B are cascaded and B has triple redundancy.

3.11. Repeat Problem 3.8 for the following conditions:
1. 2,000 non-redundant elements and 3,000 redundant elements
2. 3,000 non-redundant elements and 2,000 redundant elements
3. 4,000 non-redundant elements and 1,000 redundant elements

What conclusions can you reach from your results?

3.12. Figure 9.8 illustrates three computing systems. Which one would you recommend if reliability were the only consideration?

3.13. If the three systems illustrated in Figure 9.8 are equally effective, select the system which offers the highest reliability per dollar if the costs of the elements are as follows:

$$C_A = \$8,000 \quad C_B = \$4,000 \quad C_C = \$2,000$$

3.14. Fifteen search radar systems were tested to failure. The times to failure in hours were reported as follows: 100, 180, 200, 80, 78, 41, 420, 110, 90, 95, 115, 120, 80, 105, and 102. Calculate the mean time between failures and the standard deviation.

3.15. What decisions might be taken by the program manager if the desired reliability and maintainability cannot be reached due to budget limitations?

3.16. How might the characteristics of maintainability and mean-time-to-repair be measured?

3.17. Discuss problems associated with scheduling preventive maintenance for the following types of systems:
1. Military weapons system aboard a naval destroyer.
2. Commercial chemical process control system.
3. Commercial soda bottling plant.

Compare and justify the differences.

Chapter 4

4.1. Draw the economic flow chart of your organization.

4.2. Consider the PERT diagram for preparing engineering drawings to manufacture a receiver, as shown in Figure 9.9. Determine the following from this time model:
1. The shortest estimated time to completion.
2. The most likely estimated time.
3. The longest estimated time.
4. The "statistically estimated time."
5. The "slack time."

4.3. The PERT diagram for preparing engineering drawings to manufacture a transmitter is shown in Figure 9.10. Repeat problem 4.2 for this situation.

4.4. What is the mathematical model of your engineering organization? Where do you think it needs improvement?

4.5. Figure 9.11 illustrates the performance of an actual program. Determine the mathematical models of each organization's performance. Where do you think the performance needs improvement?

4.6. Draw the block diagram representation for the organization analyzed in Problem 4.5

4.7. Figure 9.12 illustrates an economic model showing the relationship between wages, prices, and cost of living. Observe from this model that an automatic cost of living increase results in a positive feedback loop. How can the economic process be stabilized by the addition of negative feedback loops in the form of legislative control?

4.8. The block diagram of Figure 9.13 illustrates the process of interest accrual in a saving account. Initial deposits are represented by $r(t)$; interest is represented by a constant of $A\%$ per year; total savings are represented by $c(t)$. Note that this economic model represents positive feedback.
1. Determine the relationship of total savings as a function of deposits and interest rate.
2. Determine the total savings at the end of a ten-year period for an initial deposit of $10,000 if the interest rate A is assumed to be 5% per year.

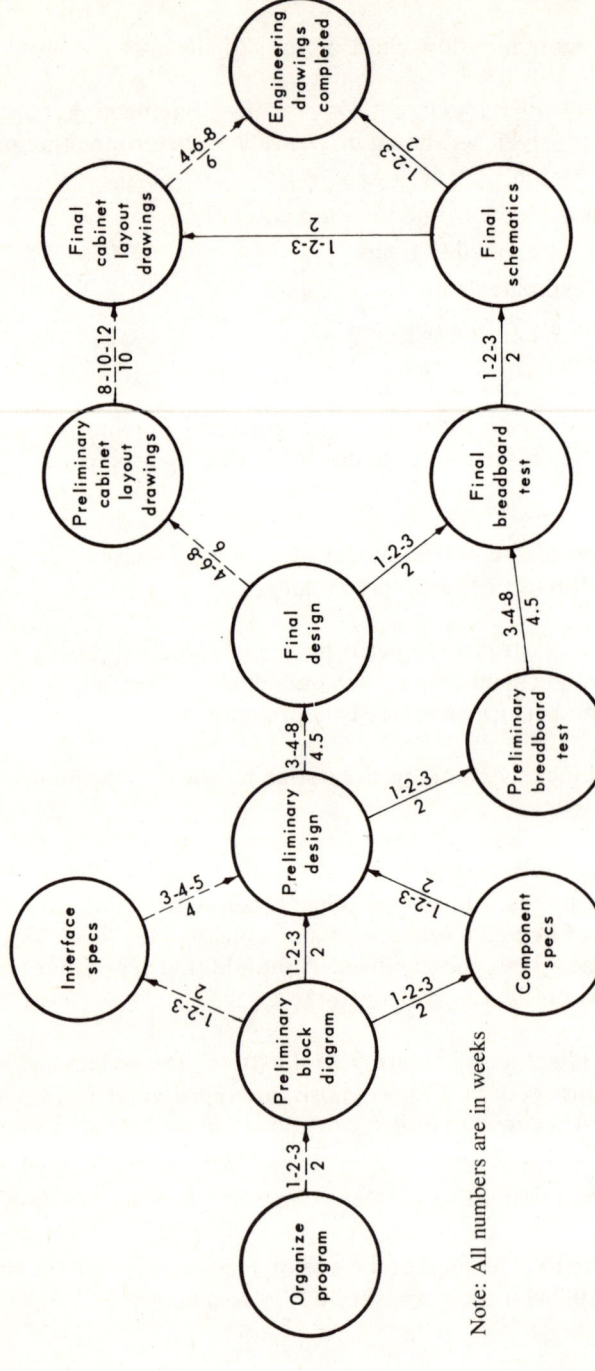

Figure 9.9. PERT Diagram for Preparing Engineering Drawings to Manufacture a Receiver

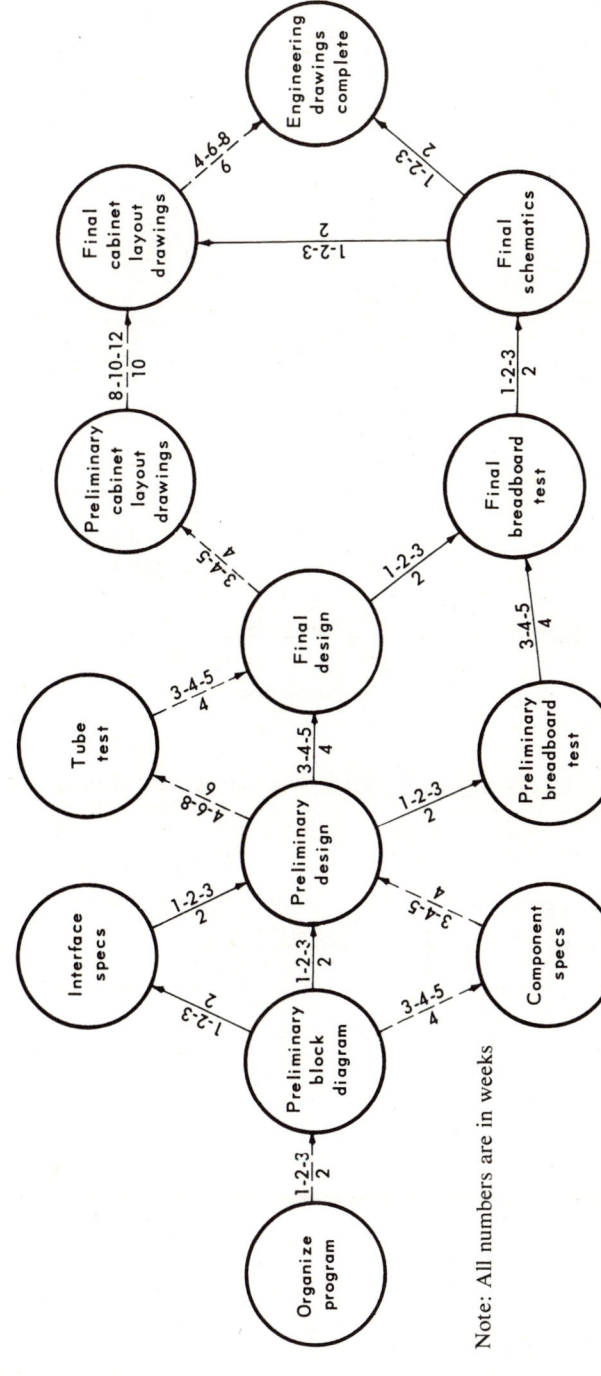

Figure 9.10. PERT Diagram for Preparing Engineering Drawings to Manufacture a Transmitter

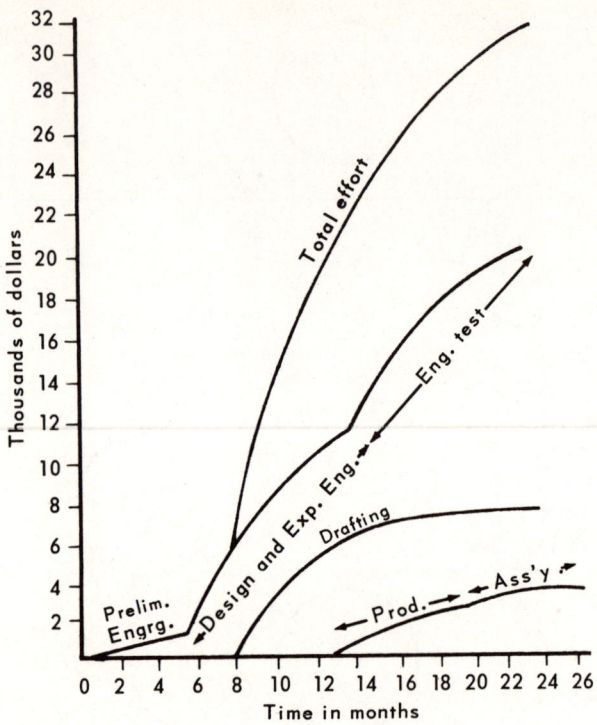

Figure 9.11. Performance Graph for an Actual Program

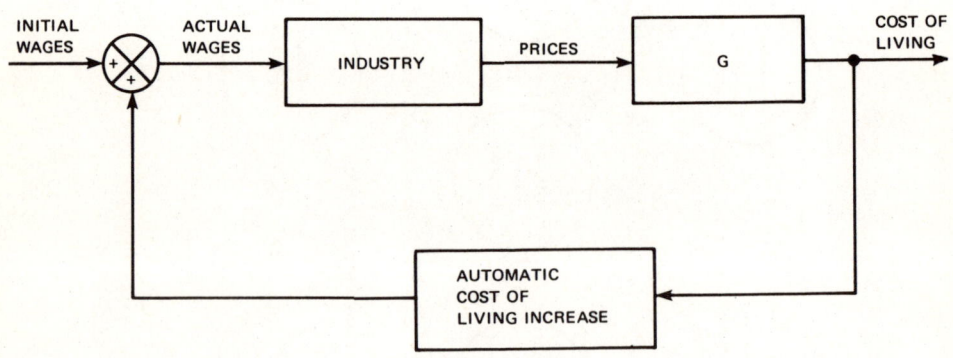

Figure 9.12. Block Diagram Showing Relationship Between Wages, Prices, and Cost of Living

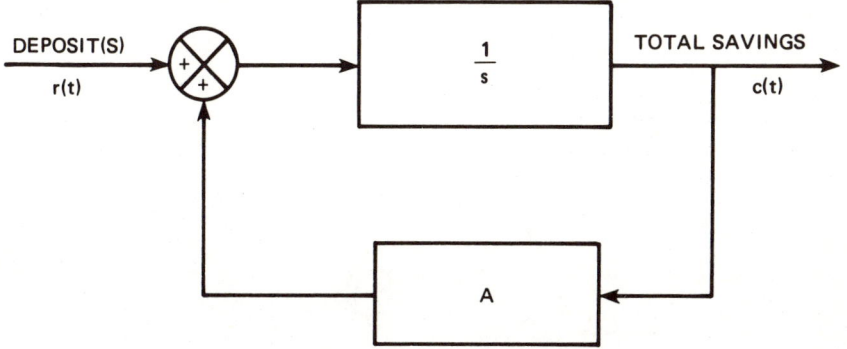

Figure 9.13. Block Diagram of Process of Interest Accrual in a Savings Account

3. Determine the total savings at the end of a ten-year period if $1,000 is deposited yearly, and totals $10,000 over this interval. Assume that the interest rate is 5% per year. How does your answer compare with part 2? What conclusions can you draw from your results?

4.9. A basic law of a free economy is the law of supply and demand. It states that the market demand for an item decreases as its price increases, and a stable market price is achieved only if the demand equals the supply. Draw an economic model of this law which contains the elements of the market, prices, supplier and demander. Assume that the input to the process is a market price change equal to zero.

4.10. Schedules developed by systems engineers usually show times at which various items of work in a major program should start and end. Why is it better to signify all of the intermediate dates than merely to show the final completion date?

4.11. The tool of mathematical modeling can be used to analyze economic problems of a much larger scope than those found in a single business organization. The economics concerned with national income, government policy on spending, private business investment, business production, taxes, and consumer spending may be represented by the block diagram of Figure 9.14. Although business production actually involves a transportation lag, e^{-Ts}, since business production lags available funds, the representation of $G_2(s)$ by the lag factor $1/(1 + Ts)$ is adequate. Assume that government policy is represented by

$$G_1(s) = A + Bs \qquad (9.3)$$

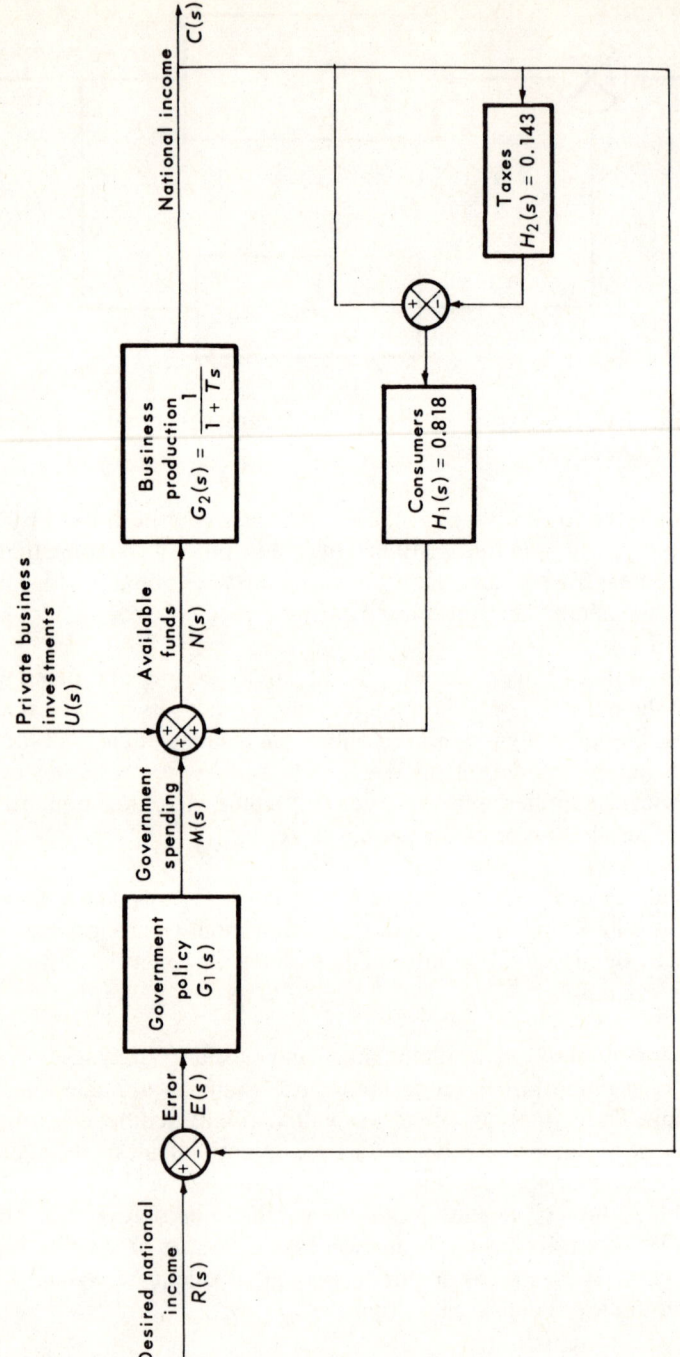

Figure 9.14. Block Diagram of National Economic Problems

which represents a lead (prediction) factor. Determine the requirements on A and B for system stability utilizing the Routh-Hurwitz criterion (see Reference 13 of Chapter 4).

Chapter 5

5.1. Develop an analog computer simulation to the following set of differential equations which represent the model of a system:

$$\ddot{A} + 1.59\dot{B} + \dot{A} + 7.31B + 0.5A = 0.02$$
$$\ddot{B} + 0.74\dot{B} + 0.49\dot{A} + 0.035B + A = 0. \quad (9.4)$$

Assume the initial conditions are zero.

5.2. Develop an analog computer simulation to the following set of differential equations which represent the model of a system:

$$\ddot{A} + 0.43\dot{B} + \dot{A} + 6.3B + 0.2A = 0.1$$
$$\ddot{B} + 0.63\dot{B} + 2.6\dot{A} + 2.2B + A = 0. \quad (9.5)$$

Assume initial conditions of A_0, $\dot{A}_0$, B_0, and $\dot{B}_0$.

5.3. Develop an analog computer simulation solution to the following set of differential equations which represent the model of a system:

$$\ddot{A} + 2.1\dot{B} + 2\dot{A} + 3B + A + C = 1$$
$$\ddot{B} + 3.1\dot{B} + 2.6\dot{A} + 2.3C + B + 6A = 0 \quad (9.6)$$
$$\ddot{C} + 6.1\dot{C} + 3.1\dot{B} + 2.8\dot{A} + 6.8C + 7.1B + 3A = 0.$$

Assume the initial conditions are zero.

5.4. Develop a logic flow diagram and write a digital program using the BASIC language to solve the following two simultaneous linear equations which represent the mathematical model of a system to be simulated:

$$Ax + By = C$$
$$Dx + Ey = F. \quad (9.7)$$

5.5. Develop a logic flow diagram and write a digital program using the BASIC language to solve the following three simultaneous linear equations which represent the mathematical model of a system to be simulated:

$$Ax + By + Cz = D$$
$$Ex + Fy + Gz = H \quad (9.8)$$
$$Jx + My + Nz = P.$$

5.6. Repeat Problem 5.4 using the FORTRAN language.

5.7. Repeat Problem 5.5 using the FORTRAN language.

5.8. Perform numerical integration on the following differential equation:

$$\ddot{x} + 2\dot{x} + x = 0. \qquad (9.9)$$

Consider the interval $0 < t < 20$ seconds and use steps of $\Delta t = 1$ second in the following series in order to obtain the required integration:

$$x(t + \Delta t) = x(t) + \Delta t \dot{x}(t) + \frac{(\Delta t)^2}{2!} \ddot{x}(t) + \ldots \qquad (9.10)$$

Assume the initial conditions are $x(0) = 1$, $\dot{x}(0) = 1$.

5.9. Repeat problem 5.8 for the following differential equation:

$$\dddot{x} + 2\ddot{x} + 3\dot{x} + x = 0. \qquad (9.11)$$

Assume the initial conditions are $x(0) = 1$, $\dot{x}(0) = 0$, $\ddot{x}(0) = 0$.

5.10. The definite integral I given by

$$I = \int_0^\infty \frac{n\,dx}{n^2 + x^2} \begin{cases} = \dfrac{\pi}{2} \text{ for } n > 0 \\ = 0 \text{ for } n = 0 \\ = \dfrac{-\pi}{2} \text{ for } n < 0 \end{cases} \qquad (9.12)$$

represents a mathematical model of a system to be simulated. Write a FORTRAN program to determine its value as a function of n.

5.11. The series resonant circuit shown in Figure 9.15 is to be analyzed on a digital computer. Draw a flow chart and write the corresponding program to compute the current i as a function of time t. From elementary electronic circuit theory, the factor $R^2 - 4L/C$ governs whether the circuit is oscillatory, critically damped, or overdamped. If this factor is less than zero, it's oscillatory; equal to zero, it's critically damped; greater than zero, it's overdamped. For the oscillatory case,

$$i = \frac{V}{\omega_n L} e^{\alpha t} \sin \omega_n t \qquad (9.13)$$

where

$$\omega_n = \sqrt{1/LC - R^2/4L^2}, \qquad \alpha = -R/2L. \qquad (9.14)$$

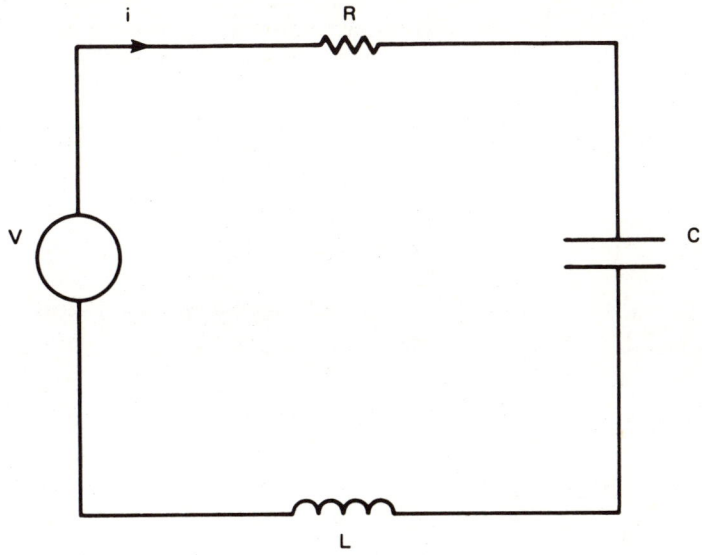

Figure 9.15. Series Resonant Circuit

For the critically damped case,

$$i = \frac{V}{L} te^{pt} \tag{9.15}$$

where

$$p = -R/2L. \tag{9.16}$$

For the overdamped case,

$$i = K_1 e^{p_1 t} + K_2 e^{p_2 t} \tag{9.17}$$

where

$$K_1 = -V[2L\sqrt{R^2/4L^2 - 1/LC}]^{-1}, \quad p_1 = -R/2L - \sqrt{R^2/4L^2 - 1/LC}$$
$$K_2 = -K_1, \quad p_2 = -R/2L + \sqrt{R^2/4L^2 - 1/LC} \tag{9.18}$$

Assume that the circuit parameters are given by $V = 10$; $L = 2.5$ henries; $C = 0.25\ \mu\text{fd}$; and that $R = 200$ for the oscillatory case; 1,000 for the critically damped case; and 2,000 for the overdamped case.

5.12. Evaluate the integral

$$I = \int_0^1 e^{-y}\, dy \tag{9.19}$$

using the Monte Carlo method and a random signal generator. How does the mean value of this integral vary as a function of the number of sample points utilized?

Chapter 6

6.1. What do you estimate your transfer function to be?

6.2. Comparing the transfer functions devised by Phillips, Tustin, McRuer and Krendel, and Bekey, which model would you propose to use in the following situations:
1. An aircraft tower controller tracking a commercial aircraft?
2. An aircraft plotter, aboard a navy attack aircraft carrier, tracking hostile and friendly targets?
3. A navy gunner aboard a fast destroyer?
4. A hunter firing at a fast-moving deer 50 yards away?

6.3. Consider the manual tracking problem where the controlled system is given by 1/s and the human being's transfer function is given by McRuer and Kerndel's model (see Equation (6.7)):

$$G_H(s) = \frac{10(1 + 0.1s)(e^{-0.2s})}{(1 + 10s)(1 + 0.1s)}. \qquad (9.20)$$

1. Determine the response in the time domain to a unit step input.
2. Determine the response in the time domain to a velocity input which can be represented as $r(t) = 10t$.

6.4. Repeat Problem 6.3 with the transportation lag decreased to 0.1 seconds. What conclusions can you reach from your results?

6.5. Repeat Problem 6.3 with the transportation lag doubled. What conclusions can you reach from your results?

6.6. Consider the manual tracking problem where Bekey's discrete model, illustrated in Figure 6.4 of Chapter 6 is appropriate to use. Assume that the controlled system can be represented by the following transfer function:

$$K_E = \frac{1}{s^2(s + 4)} \qquad (9.21)$$

The human controller's hold circuit is assumed to be a zero-order hold, his sampling period is 1 second, his gain K is unity, and his neuromuscular time constant is essentially zero. Utilize z-transformation theory analysis (See Reference 27 of Chapter 6).

1. Determine the system's response to a unit step input.
2. Determine the system's response to a unit ramp input.
3. What conclusions can you reach from your result?

6.7. Repeat Problem 6.6 when the transfer function is:

$$K_E = \frac{(s + 2)}{s(s + 4)(s + 8)} \qquad (9.22)$$

6.8. Repeat Problem 6.6 when the neuromuscular time constant is 0.1 seconds. What conclusions can you reach from your results?

6.9. Models of the human controller were recently determined, utilizing the time series method.[25,26,28] The time series analysis produces a mathematical model based on actual experimental data by reducing a waveform to white noise while identifying the correlated portion of the time series. The mathematical model is obtained by a three-stage iterative procedure based on identification, estimation, and diagnostic checking. The identification process is concerned with the generation of the series to determine a class of models that should be investigated. The estimation phase uses the data to make inferences about parameters conditioned on the sufficiency of the model chosen. Diagnostic checking analyzses the fitted model with the data in order to determine any model inadequacies and to obtain model improvements. Utilizing this approach, a compensatory manual tracking configuration, such as illustrated in Figure 6.2, was utilized to determine how well pilots could track a target on a display. For one of the pilots, his model, for 153.6 seconds of tracking data, was approximated to be:

$$G_H(s) = \frac{1.67e^{-0.2s}}{(1 + 0.33s)} \qquad (9.23)$$

when the transfer function of the controlled system, $G_S(s)$, was unity.

1. In terms of the manner in which the human controller reacts to a visually observed error, explain the source of each of these terms. Assume that the anticipation and neuromuscular time constants are zero.
2. How do the values obtained compare with McRuer and Krendel's predictions?
3. Construct an analog simulation for this system. Realize the transportation lag as a distinct block.
4. How might the transportation lag be realized in the laboratory?

6.10. When the controlled system, $G_S(s)$, in Problem 6.9 was changed from unity to a pure integration, it was found that the operator's transportation lag D was significantly decreased, while the operator's gain was significantly increased. His error-smoothing lag time constant changed very little. How would you explain these changes?

6.11. The time series method for the experiment discussed in Problem 6.9 was also used to measure the human model for the first and last 30 seconds of the experiment, which lasted for a total of 153.6 seconds. For the first 30 seconds of the experiment, the human model was determined to be approximated by

$$G_H(s) = \frac{2.5 s^{-0.2s}}{(1 + 0.5s)}. \tag{9.24}$$

Over the last 30 seconds of the experiment, the human model was determined to be approximated by

$$G_H(s) = \frac{e^{-0.2s}}{(1 + 0.2s)}. \tag{9.25}$$

Note that the model given in Problem 6.9 for the entire 153.6 seconds represents a statistical average over the length of the experiment. What conclusions can you draw from these results for the models lasting for the first and last 30 seconds? (Hint: Consider the learning aspects of the human controller over the length of the experiment.)

Chapter 7

7.1. From a trigonometric table, obtain six values of $\sin x$. Write a polynomial approximation which satisfies these values. Test your results by determining a value not included in the original six data points, but lying within the interval covered by these points.

7.2. Repeat Problem 7.1 for $\cos x$.

7.3. Discuss problems associated with automatically testing the following types of systems:

1. Military weapons system aboard a high-performance fighter aircraft.
2. Commercial chemical process control system.
3. Commercial soda bottling plant.

Would you recommend analog or digital on-line or off-line systems? Compare and justify the differences.

Chapter 8

8.1. Consider the design of an automated department store in which cash purchases are not allowed, and all payments are handled by automated billing machines with statements mailed monthly. Draw a work breakdown structure for the system, and pursue one branch of the tree down to its lowest level.

8.2. Consider the various factors involved with an automated airline reservations system. Draw a work breakdown structure for the system, and pursue one branch of the tree down to its lowest level.

8.3. Draw a work breakdown structure for the design of a system which controls large flows of surface vehicular traffic automatically via computer control of traffic lights and lane changes. Pursue one branch of the tree down to its lowest level.

8.4. Management information systems differ for different kinds of business. Compare the different kinds of information contained in the Monitoring, Demand, Triggered, and Planning Reports for the following types of organizations:
1. Commercial airline carrier.
2. Distributor of paper to newspapers.
3. Manufacturer of passenger automobiles.
4. Builder of small pleasure boats.
5. Builder of one-family private homes.
6. Military systems managing office for the design of a new fighter aircraft.

8.5. Assume that you are in disagreement with your supervisor about the objectives of your program.
1. What are the various alternatives which could help you resolve the conflict?
2. What criteria could be used for judging the obectives?

8.6. Describe the decision-making process used in your organization to fund projects for research and development. In your discussion, consider the following factors:
1. What are the bases for assigning priorities to projects?
2. How are manpower and capital resources allocated among the programs?
3. What decision process is used for choosing one project over another?

Table 9.1
Follow-On Contract Chart

Type	Contract Value	Profit (%)	Probability of Award
Spares	$5M	20	1.0
Support Equipment	$4M	10	0.8
Retrofit	$8M	16	0.75

8.7. Compare the differences in commercial and military systems engineering concerning the following:
1. Procedures.
2. Objectives.
3. Reporting.
4. Decision making.
5. Assumptions.
6. Risk assessments.

8.8. Determine the basic minimum profit of a $10M contract having a profit of 15 percent if the anticipated risk is $500K.

8.9. Determine the basic minimum profit of a $25M contract having a profit of 20 percent if the anticipated risk is $2M.

8.10. What would the overall real profit be for the situation described in Problem 8.8 if there is a 50 percent probability that a spares follow-on contract, having a value of $2M with a profit of 25 percent, might result?

8.11. What would the overall real profit be for the situation described in Problem 8.9 if there is a 100 percent probability that a spares follow-on contract, having a value of $5M with a profit of 20 percent, will result?

8.12. What would the overall real profit be for the situation described in Problem 8.9 if the follow-on contracts in Table 9.1 might result?

Index

Index

Acceleration-aided tracking, 150-153
Acceleration constant, 25
Accelerometers, 21
Acquisition, radar, 38
Activity, definition of, 76
Adaptive capability of human
   input, 142-143
   learning, 146
   task, 144-146
Adaptive systems, 23
Aided tracking, 149-153
Aided-tracking ratio, definition of, 151
Air traffic control systems, 2, 190-192
Alignment, 39-40
Along course error, definition of, 34
Alternative approaches, 179-181
Alternative solutions, 16
Amplifier, operational, 101-105
Analog computers
   accuracy of, 101
   differentiators for, 104-105
   elements of, 101-105
   function generator for, 104-105
   integrators for, 104-105
   scaling of, 101
Analysis, 14-16
   conceptual, 14
   functional, 14, 16
   mission and requirements, 14
Antenna
   boresight error, 39
   effects of wind torque on, 38, 40
Anticipation time constant of human, 137-138
Antisubmarine warfare, 191-192
Arithmetic element, computer, 112
Arithmetic mean, 30
Assemblers, computer, 117
Autocorrelation function, 127-129
Automatic test, 165-168
Availability, 45, 48, 67-68

Basic minimum profit, definition of, 182-183

Bearing wobble, error due to, 40
Beta function, definition of, 79
Binomial probability distribution, 33
Biocybernetics, 133
Biological adaptation, 142
Boresight error, 39
Braking system, railroad, 195-196
Buffon's needle experiment, 122-123

Circular probable error, definition of, 35
Combining system error, 36
Command and control systems, 190-192
Comparator failure detector, 60-63
Compensatory manual tracking, 136-137
Compilers, computer, 117
Computer
   analog, 100-108
   digital, 109-118
   hybrid, 118-120
Computer controlled testing
   analog system, 160
   digital system, 159-160
Control systems
   accuracy of, 24-28
   identification of, 21-23
   man-machine, 133-153
   test of, 165-168
   type of, 27-28
Cost
   characteristics of, 4-12
   computer control of, 95-97
   controlling, 69-97
   estimating, 93-95
Critically damped system, 88-89
Critical-path method, definition of, 76-82
Cross course error, definition of, 34
Curve fitting techniques, 161-163

Data analysis, 160-163
Data unit error, 41-42
Defect, undetected, 164
Derating, component, 51

Diagnostic testing, 159-160
Differentiators, 104-105
Digital computers
   arithmetic element for, 112
   control element for, 112
   input device for, 110
   off-line control of, 109-111
   on-line control of, 110-111
   output device for, 111
   programming of, 112-118
   storage for, 112
Dispersion of the probability density, 30
Disturbance forces, 20, 27-29
Dynamic lag errors, 20, 24-29, 33-34, 40-41

Economic flow graph, 70-72
Economic models, 190, 205-207
Encoder, error due to, 40-41
English Language Program, 167
Environmental factors, 20
Error
   bias, 20
   budget, 21, 37-42
   causes of, 37-42
   combining, 33-36
   compensation of, 28-29
   drift, 20
   dynamic lag, 20, 24-29
   predictable, 24-29, 33-34
   resultant system, 36
   steady-state, 24-29
   system, 20-21, 37-42
   system associated, 36
   unpredictable, 24, 30-36
Exponential density distribution, 33

Failure
   definition of, 46-47
   reporting, 51
   techniques for detecting, 55-58, 60-63
False alarm, 164
Feedback control theory, 2, 24-28
Feedback shift register noise generator, 126-129
First order hold circuit, 141
Fourier series, 162-163

Friction, 20
Function generator
   analog, 104-105
   random signal, 125-129
Functional analysis, 14, 16

Gaussian probability density distribution, 30
Group redundant systems, 55-63
Gyroscopes, 20

Hederodyne random signal generator, 125
Human controllers, 133-137
Human models
   adaptive, 141-149
   characteristics of, 134-135
   error-pattern-recognition, 148-149
   linear continuous, 135-139
   linear discrete, 139-140
   smoothing lag time constant of, 138
   techniques for aiding the, 149-153
   time series representation of, 211-212
   transportation lag of, 138
   unique characteristics of, 134-135
Human operator, 1-2, 132-153
Human transfer function, 132-153
Hybrid computers, 118-120
Hysteresis, 20

Identification, system, 21-23
In Service Program, 167-168
Inertial navigation systems, 21
Input adaptation, definition of, 142
Input devices, computer, 110
Integrated circuits, 53-54, 104

Japanese high-speed railway system, 194-195

Learning, human characteristics of, 142
Least-squares method, 161-162
Leveling, error due to, 40
Life expectancy of equipment, 4-12
Longest time
   characteristics of, 78-82
   definition of, 76

Machine Language Program, 167
Maintainability, 4-12, 45, 66-68
Management
   alternative considerations for, 179-181
   assumptions of, 178-179
   communications for, 184-186
   considerations of, 178
   management information systems for, 171-178
   risk assessment of, 180-184
   work breakdown structure for, 171-173
Management information systems
   description of, 171-173
   establishment of, 174-178
   function of, 174
Manual tracking
   aided, 149-152
   compensatory, 136-137
   pursuit, 136
Mean absolute error, definition of, 32
Mean time between failures, 47-48
Mean time to restore, 48
Misalignments, 20
Mission and requirements analysis, 14
Modeling
   advantages of, 99
   analog, 100-108
   digital, 109-118
   economic, 190, 205-207
   engineering organization, 86-93
   human, 132-153, 211-212
   hybrid, 118-120
   Monte Carlo, 120-124
   techniques for, 8-13
   time, 74-86
Monte Carlo method, 120-124
Most likely time
   characteristics of, 78-82
   definition of, 76
Multipliers, analog, 104-105

Naval shipboard system, test of, 165-168
Noise jitter, radar tracking, 40-41

Object program, 117

Off-line control, computer, 109-111
On-line control, computer, 109-110
On-line testing, 165-168
One sigma value, definition of, 31
Operational amplifier, 101-105
Optimization
   system, 16-17
   techniques of, 12-14
Output devices, 110
Overall real profit, definition of, 183
Overdamped system, definition of, 88-89

Parallel redundancy, 54-57
Partial redundancy, 63-66
Performance, 4-12, 19-43
PERT
   activity of, 76
   accuracy of, 85-86
   critical path of, 76
   definition of, 74-76
   event of, 76
   example of, 82-83
   slack time for, 83-85
   statistical characteristics of, 78-82
   time estimates for, 76
Poisson distribution, definition of, 32
Position constant, definition of, 25
Power consumption, 4-12
Power density spectrum, definition of, 126
Predictable errors, 24-29
Probability density function, 30
Probability of failure, definition of, 47
Probability of false alarm, 164
Probability of undetected defect, 164
Probable error, definition of, 47
Program
   damping factor of, 88
   natural frequency of, 88
   time constant of, 88
Program management, 72-74
Programming
   assemblers for, 117
   compilers for, 117
   object, 117
   source, 117
Pursuit manual tracking, 136

Quickening techniques, 149-152

Radar, 37-42
Radioactive noise generator, 125-126
Radome, 39
Railroad system, Japanese, 194-195
Random signal generators
    feedback shift register for, 126-129
    heterodyne techniques for, 125
    radioactive sources for, 125-126
Random telegraph signal, 126
Receivers, 40-41
Redundancy
    definition of, 48, 54-55
    failure detectors for, 55-58, 60-63
    group, 45-46, 55-58
    partial, 63-66
    probability analysis of, 57-60
    single element, 48
Reliability
    circuit design for, 50-51
    definition of, 45
    program orientation for, 48-51
    system, 4-12, 45-68
Reliability improvement factor, 48
Requirements allocation, system, 16
Response time, human, 138, 211-212
Resultant system error, 36
Risk Analysis Overview List, 181-182
Risk assessment
    basic minimum profit for, 182-183
    checklist for, 184-186
    guidelines for, 181-182
    overall real profit, 183
    Risk Analysis Overview List for, 181-182
    Work Breakdown Risk Analysis Summary Sheet for, 182-183
RMS error, 34
Routh-Hurwitz criterion, 207

Sampled-data system, 140-141
Schedule
    computer control of, 95-97
    controlling, 69-97
    estimating, 93-95
    PERT model of, 74-86
    system, 4-12

Semiconductor circuits, 51-54
Servo error, 20, 24-29, 33-34, 40-41
Servo noise, 40
Shortest time, definition of, 76, 78-82
Signal generators, random, 125-129
Simulation
    advantages of, 99
    analog, 100-108
    digital, 109-118
    human, 132-153
    hybrid, 118-120
    Monte Carlo, 120-124
    stochastic, 120-124
    system, 16, 99-129
Slack time, definition of, 83-85
Source program, 117
Split boundary problem, 119-121
Stable platforms, 20
Standardization, 49
Statistically expected time, 79-82
Storage, computer, 112
Structural distortion, 39
Submarine depth control system, 195-197
Synthesis, system, 17
Systems
    adaptive, 23
    design requirements for, 17
    error budget for, 21, 37-42
    error of, 20-21, 37-42
    feedback, 2, 4-7, 20-29
    human controlled, 1-2, 132-153
    inertial navigation, 21, 34-36
    man-machine control, 2, 132-153, 190-192
    models of, 8-13, 74-93, 100-120, 190, 205
    objectives of, 3-4
    optimization of, 3-4, 12-17
    performance of, 19-43
    predictable errors in, 24-29, 33-34
    procedures for engineering of, 14-17
    simulation of, 8-12, 74-93, 100-120, 190
    summary of design procedures for, 17
    synthesis of, 17
    test of, 17, 157-168

trade-off studies of, 13, 16
unpredictable errors, 24, 30-36
System associated errors, 36
Systems management
   alternative considerations for, 179-181
   assumptions of, 178-179
   communications for, 184-186
   considerations of, 178
   management information systems for, 171-178
   risk assessment of, 180-184
   work breakdown structure for, 171-173

Task adaptation, 142
Test
   analog system, 160
   computer-controlled, 159-160
   data analysis of, 160-163
   digital system, 159-160
   foundations of, 158-159
   overall plan for, 158
   reliability, 50
   uncertainties of, 163-165
Thermal noise jitter, 40-41
Thyratron noise sources, 125
Time series analysis, 211-212
Tokyo-to-Hakata railroad system, 194-195

Tolerances, 20
Tracking
   aided, 149-153
   compensatory, 136-137
   pursuit, 136
Tracking radar, 37-42, 60, 65, 133-137
Trade-offs, method of, 13, 16
Transfer functions, human, 132-153
Transistors, 52-54
Two sigma value, definition of, 31

Underdamped system, definition of, 88-89
Undetected defect, probability of, 164
Unpredictible errors, characteristics of, 30-36

Variance, definition of, 30
Velocity constant, definition of, 25

Weight, system, 4-12
Wind gust, error due to, 40
Work Breakdown Risk Analysis Summary Sheet, 182-183
Work Breakdown structure, 171-173
Worst-case design, 51

Zero-order hold circuits, 141

About the Author

Stanley M. Shinners is a Program Manager at Sperry Systems Management, Sperry Division, Sperry Rand Corporation, and is Chairman of the Systems Engineering Department of the Sperry Program for Advancing Carreers through Education (SPACE)—an industrial training program for Sperry engineers. Since graduating from the City College of New York with the B.E.E. degree and Columbia University with the M.S.E.E. degree, he has also been associated with the adjunct staffs of the New York Institute of Technology, The Cooper Union, and the Polytechnic Institute of New York. Mr. Shinners is presently also Adjunct Professor of Electrical Engineering at the New York Institute of Technology. He is author of numerous professional articles and the following books: *Modern Control System Theory and Application*, 1972; *Techniques of System Engineering*, 1967; *Control System Design*, 1964. Mr. Shinners was elected to the grade of Fellow by the Institute of Electrical and Electronics Engineers "for contributions to the theory, design and development of control systems." He is also affiliated with the American Society for Engineering Education, and the honorary fraternities Tau Beta Pi and Eta Kappa Nu. Mr. Shinners has been awarded the Professional Engineering license by the state of New York for eminence attained in engineering. His biography is listed in *Two Thousand Men of Achievement, Who's Who in the World, Who's Who in Finance and Industry, Who's Who in the East, American Men and Women of Science, Who's Who in Engineering,* and *Dictionary of International Biography.*